Experimental Film and Artists' Moving Image

Series Editors

Kim Knowles, Aberystwyth University, Aberystwyth, UK
Jonathan Walley, Department of Cinema, Denison University, Granville, OH, USA

Existing outside the boundaries of mainstream cinema, the field of experimental film and artists' moving image presents a radical challenge not only to the conventions of that cinema but also to the social and cultural norms it represents. In offering alternative ways of seeing and experiencing the world, it brings to the fore different visions and dissenting voices. In recent years, scholarship in this area has moved from a marginal to a more central position as it comes to bear upon critical topics such as medium specificity, ontology, the future of cinema, changes in cinematic exhibition and the complex interrelationships between moving image technology, aesthetics, discourses, and institutions. This book series stakes out exciting new directions for the study of alternative film practice–from the black box to the white cube, from film to digital, crossing continents and disciplines, and developing fresh theoretical insights and revised histories. Although employing the terms 'experimental film' and 'artists' moving image', we see these as interconnected practices and seek to interrogate the crossovers and spaces between different kinds of oppositional filmmaking.

We invite proposals on any aspect of non-mainstream moving image practice, which may take the form of monographs, edited collections, and artists' writings both historical and contemporary. We are interested in expanding the scope of scholarship in this area, and therefore welcome proposals with an interdisciplinary and intermedial focus, as well as studies of female and minority voices. We also particularly welcome proposals that move beyond the West, opening up space for the discussion of Latin American, African and Asian perspectives.

Elio Della Noce · Lucas Murari
Editors

Expanded Nature

Ecologies of Experimental Cinema

palgrave
macmillan

Editors
Elio Della Noce
University Toulouse Jean-Jaurès
Toulouse, France

Lucas Murari
Federal University of Rio de Janeiro
Rio de Janeiro, Brazil

ISSN 2523-7527 ISSN 2523-7535 (electronic)
Experimental Film and Artists' Moving Image
ISBN 978-3-031-70728-5 ISBN 978-3-031-70729-2 (eBook)
https://doi.org/10.1007/978-3-031-70729-2

The original submitted manuscript has been translated into English. The translation was done using artificial intelligence. A subsequent revision was performed by a third party to further refine the work and to ensure that the translation is appropriate concerning content and scientific correctness. It may, however, read stylistically different from a conventional translation.

Cover illustration: Pierre Abbeloos (standing) and Michael Snow (kneeling) with the Camera Activating Machine, near Sept-Îles, Quebec, in September 1970 during the filming of *La Région centrale*. Photograph by Joyce Wieland, all rights reserved. Courtesy of Michael Snow. © Michael Snow Studio

This Palgrave Macmillan imprint is published by the registered company Springer Nature Switzerland AG
The registered company address is: Gewerbestrasse 11, 6330 Cham, Switzerland

The title of the original French edition ("Expanded Nature – Écologies du cinéma expérimental" by Elio Della Noce and Lucas Murari) was mistakenly published in English instead of French on the copyright page in the front matter. This error has now been corrected in the updated version of the book.

Texts by Elio Della Noce, Scott MacDonald, Jean-Michel Durafour, Kim Knowles, Philip Hoffman and Janine Marchessault, Karel Doing, Chris Dymond, Alice Leroy, Rose Lowder, Chris Welsby, Yaniv Touati, Bidhan Jacobs, Lucas Murari, Teresa Castro, Colectivo Los Ingrávidos, Elizabeth A. Povinelli, Gérard Leblanc, Frédéric Brayard, Jacques Perconte, Vincent Deville, Lukas Brasiskis and Charlie Hewison.

First edition (in French)

Éditions Light Cone, Paris, 2022

Translation and review

Charlie Hewison

Scientific adviser

Vincent Deville

Editorial oversight

Emmanuel Lefrant, Eleni Gioti and Miguel Armas

Image rights

Miguel Armas

Acknowledgements

The editors of this book express their gratitude to the media historians and theorists, filmmakers and their families, film conservators and ecologists whose commitment and generosity have illuminated this project:

Nicole Brenez, Chris Welsby, Karel Doing, Leila Gharbi, Catherine Bareau, Sébastien Ronceray, Vincent Deville, yann beauvais, Philip Hoffman, Kathryn Ramey, Robert Schaller, Charlie Hewison, Éric Thouvenel, Frédérique Menant, Vicky Smith, Benjamin Cook, Mark Toscano, David Beydler, Giovanni Aloi, Claire Lasolle, Étienne Caire, Pierre Montebello, Leslie Bourgeois, Pierre Demougeot and Susanne Braun.

The organizers of the publication warmly thank the filmmakers, artists and rights holders who have kindly granted permission to publish their images in this work.

Contents

List of Contributors

Lukas Brasiskis Brooklyn College, Brooklyn, NY, USA; Columbia University, New York, NY, USA

Frédéric Brayard University of Central Lancashire, Preston, UK

Teresa Castro Sorbonne Nouvelle University, Paris, France

Vincent Deville University Paul Valéry Montpellier 3, Montpellier, France

Karel Doing Oxford, England

Jean-Michel Durafour University Aix-Marseille, Marseille, France

Chris Dymond Queen Mary University of London, London, England

Charlie Hewison Université Paris Cité, Paris, France

Philip Hoffman York University, Toronto, Canada

Colectivo Los Ingrávidos Tehuacán, Mexico

Bidhan Jacobs National School Supérieure Louis-Lumière, Saint-Denis, France

Kim Knowles Aberystwyth University, Aberystwyth, UK

Gérard Leblanc IRCAV, University Paris 3, France

Alice Leroy University Gustave-Eiffel, Champs-Sur-Marne, France; Institut Universitaire de France, Paris, France

Rose Lowder Lima, Peru

Scott MacDonald Hamilton College, Clinton, NY, USA

Janine Marchessault York University, Toronto, Canada

Lucas Murari Federal University of Rio de Janeiro, Rio de Janeiro, Brazil

Elio Della Noce University of Toulouse-Jean Jaurès, Toulouse, France

Elizabeth A. Povinelli Columbia University, New York, NY, USA

Yaniv Touati University, Montpellier, France

Chris Welsby British Film Institute, London, England

List of Figures

CHAPTER 1

Introduction: Expanded Nature - Ecologies of Experimental Cinema

Elio Della Noce and Lucas Murari

The term "Expanded Cinema" was first used by the artist Stan VanDerBeek in the spring 1966 issue of *Film Culture*. His manifesto "Culture – Intercom and Expanded Cinema",[1] accompanied by surrealist drawings, heralded the advent of a non-verbal, universal form of communication, a "film language" born of the expansion of audiovisual means. In vast *movie-dromes*, spherical domes lined with screens, humans would exchange streams of images and ideas, sounds and utopias, awakening to self-educational processes. Gene Youngblood, in his book *Expanded Cinema* (1970), noted shortly after VanDerBeek an opening of cinematic horizons[2]: the multiplication of media, technologies, and modes of exhibition and distribution of moving images would, in his view, expand the field of consciousness.

E. Della Noce (✉)
University Toulouse Jean-Jaurès, Toulouse, France
e-mail: elio.dellanoce@live.fr

L. Murari
Federal University of Rio de Janeiro, Rio de Janeiro, Brazil

E. Della Noce and L. Murari (eds.), *Expanded Nature*, Experimental Film and Artists' Moving Image,
https://doi.org/10.1007/978-3-031-70729-2_1

These "visionary" writings have in common the fact that they link two seemingly distant propositions—that of an expansion of moving images with that of an increased experience of life processes. One of the ideas that Youngblood defends in his protean and sometimes ambiguous program—the starting point of which is a technocratic bias—is the idea of "The Artist as Ecologist".[3] For the American critic and theorist, well-versed in the behaviourist psychology of his time, the awareness of such an expansion of audiovisual means opens a new field of phenomenological experimentation to the "new artist", meaning conceiving "art as environment". Expanded cinema was going to expand an experience of the living world by making possible "the simultaneous use of various media to create a total environmental experience for the audience". The relational and objectivist qualities of the expanded work would reveal "previously unrecognized relationships between existing phenomena, both physical and metaphysical".[4] The experimental works associated with the term "expanded cinema" indeed minimize the focus on narrative forms to concentrate on the system of exhibition itself and its performative powers: the projector(s), the screen(s), the beam of light, the projection space, the film and the body of the filmmaker.

> Expanded cinema isn't a movie at all: like life it's a process of becoming, man's ongoing historical drive to manifest his consciousness outside of his mind, in front of his eyes. One no longer can specialize in a single discipline and hope truthfully to express a clear picture of its relationships in the environment.[5]

Expanded Nature: Ecologies of Experimental Cinema is predicated on an ecocritical issue that Youngblood helps formulate. Doesn't this expansion of cinematic powers, related to a progressive extinction of nature,[6] lead to the imagination of *new filmic ecologies*? At the beginning of the millennium, a so-called "ontological" turn (initiated by anthropologists such as Philippe Descola and Eduardo Viveiros de Castro) sparked new philosophical debates on living beings through a critical reading of the relationships between culture and nature. The term "ecology" is associated with the field of relations between living beings and their environment (from the Greek *oîkos*, habitat; *logos*, discourse) and opening up to a wide range of uses beyond the strict framework of Western epistemology. Not only is cinema expanding, but the concept of nature itself has acquired new configurations. The ecologies of experimental cinema proposed here constitute forms of practice and activism that converse with this socio-natural diversity. They awaken a heightened sensibility toward the living world through bonds of cooperation, as well as entrust to other existents the roles of subjects and agents of filmic processes, and finally reshape the economy of film production: an ecology of perception, of the medium, of production and a multinaturalism. Comprising anthropologists, philosophers, filmmakers and artists, as well as researchers in visual studies, the *Expanded Nature* collectively converges around a common sentiment: the urgent need to *decentre human privilege*.

How does this ontological overhaul, which establishes, or restores, non-humans as animated subjects—with a point of view, a sensibility, a language—manifest itself in experimental films? How do moving images, by decentring us from our human privilege, reanimate all those living beings previously marginalized under the term of nature? *Expanded Nature* studies how so-called "experimental" filmmakers—bound together as an archipelago in a vast ocean of moving image creation—work practically and politically towards such an ecological awareness.

Ecology of Perception

Scott MacDonald's founding discussion of "ecocinema" begins with a key idea, that of experimental filmmakers often seek to restore a sensitive connection with nature through a process of "retraining of perception".[7] Armed with only a camera or film strip, these filmmakers explore the natural world, remaining open to and becoming familiar with the heterogeneity, strangeness and irregularity of living things. This ecology of attention, which can take the form of contemplation or active exchange, regenerates their perception of the living world, welcoming a diversity of subjects—animals, plants, natural entities and forces—existing in temporalities beyond the human experience. However, it would be simplistic viewing this renewed attention to nature as merely a neo-romantic or eco-responsible approach. In the context of ecofeminist practices, retraining one's perception contains the very real stakes of a struggle to reclaim the body and nature, caught up in various phallocentric oppressions.

Gestures initiated in the 1960s–1970s by filmmakers such as Joyce Wieland, Michael Snow, Chris Welsby or Larry Gottheim sowed the seeds of new attention practices: inventing new *dispositifs* for observing nature, letting them be affected by natural forces, engaging one's filming body in a symbiotic relationship with natural environments, burying the film, splicing/printing organic materials onto the film, reconstituting ecosystems during projection. The rediscovered sentiment that "[one's] life and the world's life are deeply intertwined"[8] prompts some artists to imagine links of cooperation with the living. Through craftsmanship or the engineering of filming apparatuses, these artists work to subvert the common usage and function of environmental sciences, such as biology, biosemiotics, flux detection, climatology, astronomy and so on. No longer designed for rigorous and standardized scientific evaluation, aesthetic apparatuses no longer guide but are guided by natural phenomena, taking on the role of *co-operators*. An *ecology of senses*, attentive to contingency and accidents, releases the powers to act inherent to the living, going so far as forms of "wild writing"[9] during film editing. *Expanded Nature* is a collective study of these experimental processes that unleash *the powers of action of the living world*, manifested at different scales of natural *milieus*.

Ecology of the Medium

At the beginning of the 1990s, the ecological turn coincided with another transformation, this one related to media: the digital turn. Through relations of tension and detachment, ecologies and media lead us towards what we propose to more generally term an *ecology of the cinematographic medium*.[10] The sense that the means of expression and communication were being artificialized by this transition to "all digital"—whether it be recording devices, production and distribution spaces, etc.—was quickly followed by a new wave of green, photochemical resistance among experimental filmmakers. Following the Dutch model of Studio één, the French *labos* movement emerged with the establishment of the Grenoble-based Atelier MTK in 1992. Filmmakers reclaimed the spaces, gestures, materials and machines of photochemical cinema.[11] In a critical sense, therefore, the digital turn reaffirmed a relationship with the physicality of the medium and with craftsmanship in film. The tangible nature of the medium, explored by an initial materialist movement in the 1970s—grain, textures, scratching, perforation—is largely explored anew through manual and neo-materialist practices—handmade emulsions, contact printing techniques, erosions, burials, human fluids, alchemies, etc. Film seeks to regain an intimacy with the living world, becoming a surface for exchanges between microbiological life and the creative gesture.

However, even though a major part of the book *Expanded Nature* is devoted to the renewal of photochemical practices and seeks to explore their organic constellations, we felt it appropriate to include practices in other media, and especially to open the discussion to the field of new media. Through the tensions of this dual digital and ecological turn, it seemed to us that the entire relationship between visual artist and medium of expression was being reconfigured. The digital medium seems caught up in this transformation, seemingly conferring on it the task of expressing another "ecology", which we would call a *subtractive ecology*. This ecology is less about the physicality of the material than a rupture with the real: an exploration of the surreal nature of algorithms, the kinetic and colorimetric superpowers of the screen, the connective vulnerability to diffusion, and, conversely, the environmental value of the medium itself. *Expanded Nature* thus proposes to question the ecological potentialities of different cinematic media, and investigate their unique capacities to reveal the infrastructures of the living world.

Ecology of Production

> We are not in an economy but an ecology, a grassroots network, filmmakers helping each other, outside of the capitalist system. Furthermore, we don't work with "images", but with organic, physical material that comes from the earth: salts, silvers, minerals.[12]

In this statement by filmmaker Pip Chodorov, the idea of an "ecology" as distinct from an "economy" of production is understood in its political sense, highlighting the increasing tendency among these independent filmmakers to leave the dominant circles of the audiovisual industry (where money remains inscribed on the front and back of images). Since the 1960s and the founding of the very first co-operatives, experimental filmmakers have been devising non-profit and sustainable ways of organizing, transmitting, sharing, creating, distributing and preserving film. Sixty-eight artist-run film labs[13] now welcome all filmmakers, beginners or seasoned veterans, and are structured on the basis of volunteering, knowledge sharing, DIY,[14] the reuse of machines and the self-empowerment of processes. This includes the manipulation of cameras, materials and emulsions, as well as machines such as editing tables, printers, developers and projectors. In Dominique Noguez's seminal text *Éloge du cinéma expérimental* (1979) (*In Praise of Experimental Cinema*), he foresaw a DIY and "somewhat ecological" tendency of experimental filmmakers, emphasizing the maximum and necessary limitations to producing a film.[15]

In these independent labs [*labos autogérés*], of which France has the highest number, ecofriendly communities are emerging. They organize workshops and bring together filmmakers and the curious, witches and alchemists, to experiment with alternative processes such as phytography[16] or ecoprocessing.[17] From this experimental soil, Film Farms emerged: creative residences in rural settings driven by alternative philosophies,[18] as well as retreats and expeditions in natural environments.[19] Like mycelium, these multiple initiatives form an interconnected network of underground filaments, a "grassroots network" that grows slowly in the shadow of the industry or large institutions, ultimately changing our vision of duration and the time it takes to produce a film.

From Multiculturalism to Multinaturalism

The subjectivity that reigned over modern philosophy was heir to the Cartesian *cogito*; in other words, the existence of the subject was deduced from its ability to think. Through this postulate, Descartes distanced humans from animals and the rest of the living world, valuing above all a thinking and rational subject. The recent ontological turn precisely questions anthropocentrism's claim to primacy. As early as the 1940s, experimental filmmakers had embarked on this long path of decentring. Through visual ethnography and fieldwork, they sought to establish alliances with other modes of thought and existence (Maya Deren and Gregory Bateson, Jean Rouch, Raymonde Carasco, Trinh T. Minh Ha). In recent years, the Brazilian anthropologist Eduardo Viveiros coined the neologism "multinaturalism"[20] to address these new epistemic challenges. This conceptual problematization, in a certain way, updates the idea of "multiculturalism". Based on the ontologies of Amerindian cultures, Viveiros de Castro proposes an inversion: a single culture, multiple natures. "If Western multiculturalism is relativism as public policy, then Amerindian perspectivist shamanism is multinaturalism as cosmic politics."[21]

The opening of *Expanded Nature* to this ontological turn leads to considering film experimentation as a purveyor of other modes of existence and relationships: a *multinaturalist dispositif*. It is no longer a single "nature", unified and external, but a multiplicity of natures and cosmological subjects that the filmmaker welcomes and engages in conversation on distinct cosmic levels. The words of several Indigenous filmmakers (Amerindians, Zapotec, Mixtec, Aboriginal), rooted in animist or totemist ontologies—here put into conversation with those of anthropologists or researchers in visual ethnography—set us on this path. These exchanges reveal a *trans-species power* of film to play out the metamorphoses, the passages from one bodily envelope to another, the journeys of the spirit to ancestral myths: being animal or plant, transforming into river or ritual object, and thinking as a tree or mountain. The different works discussed in *Expanded Nature* strive to invent forms of connection to the world that foster new creative possibilities and engage the imagination in new ways of seeing, hearing, feeling and perceiving cosmological heterogeneity. The proposal of such decolonization therefore necessitates transcending exceptionalism centred on a single species.

Multifocal Approaches

To accompany the readers/explorers towards this idea of a *decentring through film*, *Expanded Nature* evolves in the manner of a multifocal lens. In a slow speculative zoom-out movement, from the micro to the macroscopic, it invites the reader to see the filmic apparatus in terms of the different scales of its ecological singularities: from the most minute organic materiality to the reconstituted environments of "exhibited cinema".

The first Part, "Nature Films/Film Natures", posits the ecology of medium at the most elementary scale. It postulates that the fundamentally ecological properties of film are *perceptual* and *material* and seeks in particular to examine them at the heart of photochemical practices. Elio Della Noce draws on psychologist J. J. Gibson's ecological theory of perception, which he applies to environmental observational devices in the history of structural cinema. The author's aim is to bring the question of cinematic perception back to the inaugural moment of filming and thus remove it from the projective (and indirect) situation of the cinema theatre. This idea of *film as an act of direct perception* is reaffirmed by Scott MacDonald, who proposes a genealogy of ecocinematic long takes from Larry Gottheim to the Sensory Ethnography Lab (SEL), via Sharon Lockart and James Benning. MacDonald deepens his canonical hypothesis of retraining perception through the motif of the long-shot film. In an even more decentred gesture, Jean-Michel Durafour leads cinematic perception towards its entomological qualities, scrutinizing in the filmmaker Paolo Gioli, among others, an analogy between the shutter and the fluttering of a butterfly's wings. The materiality of film is then discussed, through the prism of the new materialism of Jane Bennet or Karen Barad. These two philosophers open the ontological problem of matter to its instability and its queerness.[22] In Kim Knowles' text, the haptic commitments of filmmakers in what the media theorist and historian calls a "renewed approach to materialist film" are studied through the environmental "contact" practices of filmmakers Emmanuel Lefrant and David Gatten, the handmade emulsion researches of Esther Urlus and in the "body fluid" films of Vicky Smith. In Philip Hoffman and Janine Marchessault's chapter, the heterogeneity of matter meets the question of process and ecofeminism, from Joyce Wieland's craftsmanship to the filmmakers of the Film Farm—Jennifer Reeves, Kelly Egan—and the Indigenous collective Odeimin Runners Club. Finally, filmmaker Karel Doing advocates for a disorderly and multi-species cinema, starting from an anarchic, recalcitrant materiality, of unstable identity, embodied in his films through vegetal and silty variations.

In Part II, "Cooperating With The Living", we gradually leave the medium scale to reveal the mesological scale[23] of the filmmaker's relations with the living world (environments, forces, flows, plants, animals). At the scale of the terrestrial environment, Chris Dymond follows up on Jean-Michel Durafour's idea of "seeing in the place of a butterfly" by introducing the idea of "filming like a plant": from Anna Atkins' Prussian blue algae to Miles McKane's fungal visions, to Karel Doing's phytosemiotic systems, Dymond invites the plant to exceed the frame that natural history had assigned to it. Alice Leroy in turn focuses on an intriguing fact of cine-natural history. According to her, "the technological history of cinema overlaps with that of the destruction of natural worlds", which leads her to visit experimental films as moors traversed by endangered animals. This is discussed by filmmaker Rose Lowder, as she films poppies, salt marshes and turtles at the edges of socio-natural environments, developing techniques of in-camera editing that resemble the act of picking and arranging a bouquet. Next comes an autobiographical text, leading to the celestial realm of filmmaker Chris Welsby. He recounts his awakening to an ecological consciousness (as a hiker and sailor), then returns to his pioneering filmic apparatuses oriented by climatic, homeostatic and astronomical forces (tides, cycles of the day or equinoxes). Finally, two studies take Jean Epstein's insular films as their starting point. The phenomenological poetry of these films opens a dialogue around environmental sciences applied to filming systems, here considered through the phenomenology of clouds or the detection of terrestrial fluids (smoke, gas, mirages). Yaniv Touati focuses his study on French filmmakers Philippe Cote and David Dudouit, who survey oceanic cloud phenomena, while engineer in technology Bidhan Jacobs traces the genealogy of the filmic systems used to detect stochastic properties of fluids. Following a history of the evolution of media, cameras, optics, interfaces and broadcast screens, he draws an atmospheric connection between the photochemical vapours of Jean Epstein, the video art of Leighton Pierce or Bill Viola, and the digital volutes of Robert Cahen or Jacques Perconte.

Part III, "A Decentred Cinema", takes a further step back, seeking to widen the speculative frame. It expands its outlook to philosophical and anthropological scales that help reshape the meaning of the word "nature", in dialogue with different ways of apprehending the world, including those among Indigenous communities or peasant farmers. These different approaches and testimonies nourish the idea of film as *multinaturalist dispositif*. The first two chapters address a fundamental question in anthropology and cinema: animism. Lucas Murari focuses his analysis on the feature film *The Origin of the Night (Amazonian Cosmos)* (1973–1977) by Lothar Baumgarten. The film was based on an Amazonian Tupi myth discussed by Claude Lévi-Strauss in his *Mythologiques* series. Shot in a forest by the German artist, it seeks to highlight the complexity of multinaturalist cosmologies present in the thought of Amerindian cultures. Teresa Castro directs the debate on animism in another direction. Cinema, which offers attention to infinitesimal existences, is in her view an invitation to discover non-humans in all their vitality: "To *reanimate*, and not to animate, nature – to revive, give life and vitality to those who have lost them, those from whom such qualities were taken away". Her study, based on the ecofeminist philosophy of Val Plumwood, moves from a pioneering work of the Lumière brothers, *Repas de bébé* (*Baby's Meal*, 1895) to the works of experimental filmmakers Charlotte Pryce and Martine Rousset. The text that follows is a manifesto written by the Colectivo Los Ingrávidos, a group formed in Oaxaca in Mexico in 2012, which has produced more than 200 films over the last decade. In "Shamanic materialism: 77 theses on the audiovisual", the collective presents, in a literary yet experimental form, a radical and (cosmo)political alternative in the way we make cinema and audiovisual works. The chapter by anthropologist Elizabeth A. Povinelli addresses the work of the Karrabing Film Collective, a group of over 50 Indigenous people from the Northern Territory of Australia, which she helped found and actively collaborates with. The discussion here recounts the story of the Aboriginal collective and its geopolitical, cultural, existential and mythical challenges "in the era of late colonial liberalism". The reflection proposed by Gérard Leblanc also invites a change of perspective on nature, starting from documentary procedures that "listen" to peasant farmers' voices. Three films by Gérard Leblanc, co-directed with Catherine Guéneau, examine the potential of organic farming in the face of the agri-food industry. Philosopher Frédéric Brayard focuses on the works of Jacques Perconte, discussing formal strategies based on digital compression programs. His reading of Perconte's work, via Tristan Garcia's speculative realism, leads to "films after nature" marked by the equality of being and the agency of all things.

The fourth and final Part, "Expanding the Experimental Field", concludes this slow zoom-out. Ultimately, it seeks to provide an overview of the various media and procedures that reveal "nature" reactivated in other contexts: exhibited cinema. The first text is an original interview with visual artist Jacques Perconte conducted by Elio Della Noce. The conversation follows the artist's journey from his first shots in/of nature—rivers, mountains, seas, birds—to the exhibition of his works in contemporary art spaces. It focuses on the ecomediatic specificities of digital technology, whose multiple infrastructures are explored by Perconte. Next, Vincent Deville's study is based on the observation of a decentring gesture in cinema, which the author calls a "leap into nature" and which he extends all the way to the ecoprocessing baths practised in French artist-run labs. The article maps this gesture across the history of cinema, from Jean Epstein's *Mauprat* (1926) to the "leaps" of Ralph Steiner, Miles McKane, Helena Gouveia Monteiro or filmmakers from the MTK collective in Grenoble. Lukas Brasiskis investigates the "techno-natural" representations imagined by contemporary artists whose primary medium of expression is the moving image. He situates this eco-aesthetic and hybrid debate more precisely within the theme of the Anthropocene as addressed by new technologies, discussing works by Chia Wei Hsu, Pedro Neves Marques, Emilija Škarnulytė and Jonathas de Andrade. Charlie Hewison, on the other hand, looks at the performances of artists from the French experimental scene: the Nominoë Collective, Silvi Simon and Gaëlle Rouard. According to Hewison, their works propose an "ecopoetics of re-enchantment" through non-alienated, artisanal and collective strategies of artistic creation, more broadly inscribed in a French history of artist-run film laboratories.

Notes

1. Stan VanDerBeek, "Culture – Intercom & Expanded Cinema", *Film Culture,* no. 40, Spring 1966.
2. Gene Youngblood, *Expanded Cinema,* New York, P. Dutton & Co., 1970.
3. Ibid., p. 348.
4. Ibid., p. 346.
5. Ibid., p. 41.
6. The WWF (World Wide Fund for Nature), an NGO founded in 1961 by four British biologists with the aim of raising funds to protect wildlife and its habitat, estimates that 68% of animal species have become extinct since 1970. The WWF's 2022 *Living Planet Report* is entirely downloadable from this address: https://living planet.panda.org/en-US/. A study cross-referenced by dozens of institutions and NGOs on possible viable actions to address this mass extinction was published in *Nature* in 2020: David Leclère, Michael Obersteiner et al., "Bending the curve of terrestrial biodiversity needs an integrated strategy", *Nature,* no. 585, September 2020, pp. 551–556.
7. Scott MacDonald, "Toward an Eco-Cinema", *ISLE: Interdisciplinary Studies in Literature and Environment,* Vol. 11, no. 2, Summer 2004, p. 109.
8. David Abram, *The Spell of the Sensuous, Perception and Language in a More-Than-Human World,* New York, Vintage Books, 1996, p. 29.
9. See here the concept of "wild writing" as defined by Californian poet Gary Snyder in his ecopoetic manifesto "Unnatural Writing", taken from a conference at Squaw Valley, California, "Art of the Wild", July 1992. Gary Snyder, *A Place in Space. Ethics, Aesthetics, and Watersheds,* Washington DC, Counterpoint, 1995, pp. 163–172.

10. See the discussion on obsolescence in the era of new media in Kim Knowles, "The Matter of Media", *Experimental Film and Photochemical Practices*, Basingstoke: Palgrave Macmillan, 2020, pp. 1–24; see also Scott MacKenzie and Janine Marchessault (eds), *Process Cinema: Handmade Film in the Digital Age*, Ontario, Queen's University Press, 2019.
11. For more details on the history of independent European and French laboratories see: Pip Chodorov, "The Artist-Run Film Labs", *Millennium Film Journal*, Vol. 60, Autumn 2014, pp. 28–36; also Charlie Hewison, "Révèler les traces, conversation potentielle avec Emmanuel Lefrant, Frédérique Menant, Olivier Fouchard et Mahine Rouhi", in *Débordements*, "Terrestres après tout", no. 2, 2020, pp. 211–236.
12. Pip Chodorov, "The Artist-Run Film Labs", *art. cit.*, p. 36.
13. See the *Film Labs* site, updated at the end of 2021, which offers a planisphere of the current network, a film catalogue, the latest distribution news and a set of didactic tools. https://www.filmlabs.org/fr/ [accessed 28 May 2024].
14. DIY [Do It Yourself], a term commonly used to refer to inventive, autonomous and *lo-fi* craft practices.
15. Dominique Noguez, *Éloge du cinéma expérimental* [2nd ed.], Paris, Paris Expérimental, 1999, p. 27.
16. Phytography is a contact technique invented and developed by filmmaker Karel Doing since 2016. It involves, after immersing plants in an environmentally neutral photosensitive solution (based on vitamin C, baking soda and cold water), placing them in contact with film or photosensitive paper. The plants, exposed to natural light for varying times, produce hyper-detailed negative prints on the silver surface. https://kareldoing.net/phy/Phytography.html [accessed 12 November 2021].
17. On this issue of *ecoprocessing* see the wild recipe book by Kathryn Ramey and Dagie Brundert. They have recorded the recent evolution of alternative and organic soups used for development, tinting and toning: from the initial invention named "Caffenol" to their own experiments with wine, fruits and vegetables. In Kathryn Ramey, "In the Soup 3: Eco-hand processing", *Experimental Filmmaking: Break the Machine*, New York and London, Focal Press, 2016, pp. 197–216. Also see the *GREEN Workshops* given at the *Film Farm*

by Philip Hoffman, Ricardo Leite and again Karel Doing and Dagie Brundert. https://philiphoffman.ca/category/process-cinema/process-cinema-other-workshops-film-farm/ [accessed 8 November 2021].

18. The first *Film Farm* is that of Philip Hoffman, which he founded with Marian McMahon during the summer of 1994 in Mount Forest, Ontario. Then comes the *Allerdale Film Farm* (2017) and the *Analogue Farm* (2018) established in England.
19. The *Wilderness Filmmaking Expedition* by Robert Schaller offers a stay in the Rocky Mountains that introduces a "zero emission" cinema, while the aptly named *Grazing not Gazing* by Vicky Smith invites you to imitate cows by scrutinizing grass with a macro lens. For more details on these ecofilmic retreats see the respective sites of: the *Film Farm* by Philip Hoffman, https://philiphoffman.ca/film-farm/; the *Wilderness Filmmaking Expedition* by Robert Schaller, http://www.handmadefilm.org/wildernessFilmmaking/; *Grazing not Gazing* by Vicky Smith, https://vickysmith.blog/2020/01/19/341/ [consulted on 12 November 2021].
20. Eduardo Viveiros de Castro, *Cannibal Metaphysics. For A Post-Structural Anthropology* [2009], trans. Peter Skafish, Minneapolis, Univocal Publishing, 2014, pp. 65–75.
21. Eduardo Viveiros de Castro, "Cosmological Deixis and Amerindian Perspectivism", *The Journal of the Royal Anthropological Institute*, Vol. 4, no. 3, September, 1998, p. 472.
22. See Jane Bennett, *Vibrant Matter: A Political Ecology of Things*, Durham and London, Duke University Press, 2010; and Karen Barad, "Nature's Queer Performativity", *Qui Parle*, Vol. 19, no. 2, Spring/Summer 2011, pp. 121–158.
23. Mesology is an interdisciplinary science that studies the relationship of living beings with their living environment. See Augustin Berque, *Poétique de la Terre: Histoire naturelle et histoire humaine, essai de mésologie*, Paris, Belin, 2014.

PART I

Nature Films/Film Natures

CHAPTER 2

Film as an Act of Ecological Perception

Elio Della Noce

> An image of the sea breaking on the shore is not so much a symbol of passions, rather the reality evoked by passions simply reinforces this simple and non-transferable statement of the sea by the sea.[1]

Ecological Perception

One of the acts common to all human experience and which connects our body to the natural environment is that of visual perception. In this initiation to the outside, not only our eyes, but also our head, shoulders, hands, feet and nascent movements are invited to an ecological feast of the visible, taking the form of exploration. This perceptive interaction of the human organism with the natural environment is something that our ***more-than-human*** modernity was supposed to distance us from: it was supposed to substitute the terrestrial horizon with a verticality, feed our attention with artificial signals and regulate our movements with

E. Della Noce (✉)
University Toulouse Jean-Jaurès, Toulouse, France
e-mail: elio.dellanoce@live.fr

E. Della Noce and L. Murari (eds.), *Expanded Nature*, Experimental Film and Artists' Moving Image,
https://doi.org/10.1007/978-3-031-70729-2_2

pendular mobility. Perception was supposed to entail the futuristic game of effervescences, urban flickering, epilepsies, apparent stimulations, until we were left wanting. Here is today's new *lament*: what is there left to perceive? An *ecology of perception*, perhaps this is what a certain cinema invites us to. Perceiving the pine needles of a forest, the water that rushes down the slope of a river, the wild grasses under our feet, a cloud laden with rain, the effects of the wind on the sea—far from any "spectacular" impression—informs us of our real belonging to nature. It seems that *the act of filming*, a decisive step in the filmic event, could be considered as a reactivation of this "ecological" dimension of visual perception. This is what our study seeks to discover.

But what exactly is this *ecological perception*? The American perception psychologist James Jerome Gibson published *The Ecological Approach to Visual Perception*[2] in 1979. Upon its release, the ecological theory of visual perception caused a major upheaval that shook the foundations of the discipline. The psychology of perception had until then indeed consolidated around a major trend known as the "inferential approach". Hermann Von Helmholtz, a nineteenth-century Prussian scientist, had introduced the idea of "unconscious inference": to perceive, according to Von Helmholtz, was to call upon prior knowledge, stored in memory in the form of representations, in order to interpret data from our sensory systems.[3] Cognitive psychology, which dominated studies on the subject from the twentieth century onwards, took up this idea of inference to extend its theories, whether by a psychologist like Irvin Rock or the neuropsychologist Richard Gregory. In the early 1980s, the neuroscientist David Marrl structured his "computational" theory on a similar basis, describing the visual system as a vast computer program, composed of algorithms, which structures the environment on the basis of information contained in the photoreceptors.[4] James J. Gibson's ecological approach differs radically from inferential approaches, as we'll see.

Firstly, Gibson addresses visual perception as "direct" perception and no longer "indirect" as was the case with inferential theory.[5] We move away from a laboratory mindset where perception, studied in an anthropocentric, closed and idealistic observation context, was a matter of mental representation. For Gibson, "direct" perception is instead the activity of discovering meaningful objects in the environment by moving around in it.[6] It can only be understood in its mutual relationship with action. Such an exploratory activity brings the organism into interaction with the natural phenomena and living beings, human or otherwise, that surround it. Indeed, and this is the second point, Gibson opens the study of perception to an extended animal/environment system, where ecological qualities become just as essential as the internal qualities of the perceiving animal (including humans). The third cardinal point of Gibson's theory lies in the notion of "reciprocity" of this animal/environment system, which he formalizes in a very audacious proposition. For our psychologist, visual perception indeed relies on "information", but this is data that the perceiving subject *picks up* from the surrounding world. It is no longer inferences stored in memory or algorithms to compute but information *contained* in the environment in an objective form and which are *picked up* by the exploratory activity, thanks to what Gibson calls "the ambient optic array". However, one thing particularly catches our attention: the last chapter of the book intriguingly focuses on moving images and explicitly on cinematographic art. Here Gibson begins a theory of montage and then reception through a study of "progressive pictures" as a possible application of the ecological qualities of human perception.[7]

Returning to this idea of visual perception as an act of *exploration* and *ecological reciprocity* indeed leads us to reconsider film theory. Since the first avant-gardes and the exaltation of poet-filmmakers, cinema has been the subject of scholarly interest for questions of perception, sublimated into a kind of futuristic reverie. However, ahead of being a machine that "extends the field of perception", a "superhuman eye" or an "intelligent machine",[8] isn't cinema simply a means of investigating the *ecological value of perception*? For filmmakers, doesn't ecological perception—this relationship they see themselves weaving with natural light, its variations and its impact on the encountered obstacle, on the path to take, leading to the rock, to the fluctuations of the wind in the trees—seek an immediate translation in the act of filming? Cinematographic theories of visual perception remain very much rooted in a *Gestalt*-influenced psychology of perception, associating the physiological mechanisms of human perception with the later stages of the filmic event: those of projection and reception in the darkened theatre.[9] Yet, unlike Gibson, it seems more productive to resituate the question of visual perception within the initial experience of filming. Contemporary historians of experimental and avant-garde cinema lead us down this path and, by focusing on the environmental practice of filmmakers, open the question of cinematic perception to current ecocritical issues. According to Kim Knowles, "what we perceive and how we perceive it"[10] are new forms of resistance located at the heart of "the ethics of the image", and contribute to the critique of the *New Climate Regime*.[11] For Scott MacDonald, *ecocinema* works towards a "retraining of perception"[12]: it offers an "alternative to conventional media-spectatorship", "an 'Edenic' respite from conventional consumerism"[13] through the invention of other modes of attention to nature. We understand that today the question of *perceiving through film* involves interwoven aesthetic and ecological implications and that it raises broader questions about the politics of representation.

Ecology and Structural Cinema

The emergence of an ecological consciousness in film practice is a relatively late development, originating with some filmmakers from the late 1960s and early 1970s. Experimental practices stemming from so-called "structural"[14] cinema began to place visual perception of the natural environment at the centre of the filmic process, notably through "controlled observational device[s]"[15] where the camera's multiple recording functions are manipulated, even diverted: in-camera editing, playing with the shutter and exposure times, time-lapse effects, cameras oriented mechanically or by natural forces. This environmental constant emerged within American, Canadian and European structural cinema, but particularly in the British avant-garde of the London Film-Makers' Co-operative,[16] with young filmmakers like Chris Welsby, William Raban and John Woodman. In 1976, the critic and historian of experimental cinema Deke Dusinberre proposed naming this general trend *Landscape Film*.[17] What appears in this new aesthetic is the proposition of a *perceptive analogy* between the "direct" perception of the filmmaker (at the time of shooting) and the produced filmic object: the filmmaker shows the way he operates via a *co-perception* of the landscape and the method of recording it, he places at the heart of the content and form of the film the aleatory sensuality of the landscape and the forces that run through it,[18] and seeks an objective duration when editing (the films follow the natural cycles of the sun, night, seasons, equinoxes, etc.). Using the cinematic properties of the landscape, the film now studies perception as a fact of ecological observation, a phenomenon that refrains from any interiorization. The structural filmmaker thus tends towards an aesthetic objectivity where the act of perception is re-enacted in the nudity of its ecological components. In this way, the structural film becomes an instrument for studying what Gibson referred to as an *ecological exteriority* inherent to our visual perception.

To grasp the origins of this ecological trend in structural cinema we must first take a few steps back and return to its founding film: *3/60 Bäume im Herbst* (*Trees in Autumn*, 1960) by the Vienna School filmmaker, Kurt Kren. This film, made using in-camera editing, features a succession of shots of tree canopies taken in the vicinity of Vienna in autumn. Shots of bare trees, at different scales and camera angles, follow one another at predetermined frame rates. These rates replay a "score" of lines and dots hand-drawn by Kurt Kren. According to the English filmmaker and theorist Malcolm Le Grice, this is the first "structuralist"[19] film. In saying this, Le Grice was distinguishing himself from the "structural" label upheld by Paul Adams Sitney and Peter Gidal: "Its structuralism is the result of the application of system, not to subsequent montage of material already filmed with an unrestrained subjectivity, but to the act and event of filming itself". He continues: "Objects are seen as an amalgam with their space and more especially with their time, as the process of their accessibility through acts of perception".[20] Le Grice emphasizes this dual idea of structuralist film as an "act of filming itself" and as a process of accessibility of objects "through the act of perception". The "ecological" nature of the filmic apparatus—a filmmaker in perceptive relation with natural objects (trees)—seems to bring cinema back to its ontology. The idea of "natural environment" meets that of the filmic "medium" (apparatus, material), a medium that acts as an intermediary between perception and the filmmaker's natural environment. *3/60 Bäume im Herbst* is based on this new structural process and invents a phenomenological object (the film), by linking the observer's space/time with the space/time of non-human living beings. This birth of "structuralist" cinema, linked to the successive perception of tree branches, certainly raises questions. *3/60 Bäume im Herbst* inaugurates new procedural and systemic approaches[21] for filmmakers, with an increased focus on the perception of nature. This focus will lead to explorations of the diversity of interactions between observation techniques and the immanence of nature. From this source, or let's say, from these roots, we see that the question of perception as an "act" brings together both the ecological approach of psychologist Gibson and Le Grice's film thinking. In what way is film, as an act of filming, an ecological act of perception? How does this act of perception lead to deepening our knowledge and experience of the natural environment?

Visual Kinesthesis

In Gibson's theory, one of the fundamental points of perception as *information pickup* is the idea of *visual kinesthesis*. In his own words: "The purpose of vision, I shall argue, is to be aware of the surroundings, the ambient environment, not merely of the field in front of the eyes".[22] This awareness of the totality of the environment, according to Gibson, is achieved in the animal by an information pick-up that specifies three movements: rotation of the head relative to the body, the scanning or *panning* movement; locomotion relative to the environment or *tracking*; and the movement of the limbs relative to the body, protrusion and particularly *manipulation*. Let's see how the *visual kinesthesis* at work in our direct perception finds a fruitful application in the observation procedures of landscape film.

A prototype of a remote-controlled tripod, about the size of a man, is placed on top of a high mountain, in a desolate region of Quebec, north of Sept-Îles. Attached to the mechanical tripod is a 16 mm camera connected to a sine wave generator. Programmed frequencies activate 360° panoramic movements, but also zooms and rotations on the horizontal plane, at variable speeds and multi-axial scanning. Malcolm Le Grice saw the Canadian filmmaker Michael Snow as a direct heir to the structuralism of *3/60 Bäume im Herbst*. His film *La Région centrale* (1971) is a veritable demonstration of visual kinesthesis. Here, it is the *scanning* function of visual perception that is systematized by the apparatus. The general idea of a "natural environment" seems to be condensed in the chosen location: a mountain summit that offers an all-encompassing view of the wild moorland and distant landforms, at the edge of the sky. Michael Snow tells us: "I wanted to make a film in which what the camera-eye did in the space would be completely appropriate to what it saw, and at the same time equivalent to it".[23] As well as reproducing the unity of method and subject characteristic of structuralist activity, he highlights the objectivist dimension of the experience. Indeed, *La Région centrale* differs from his previous controlled observational devices (notably from *Back and Forth* [1969] and indoor devices), in that by radicalizing the landscape subject and the duration (180 minutes), the machine allows him to extend the amplitude of the camera's panoramic movements to a sort of infinity. But the *dispositif* is ambiguous. This machine, non-human therefore, is nevertheless designed on a human scale. It broadens the view of the landscape from the height of the filmer.

La Région centrale leads us to several hypotheses: is perception based on this kinaesthetic ability to embrace the entire visible environmental field? Is cinema a means of studying the cosmic dimension of human perception? This awareness of the totality of the environment in the act of direct perception is here reactivated, embodied by the cinematographic apparatus.

In the 1980s, the English filmmaker Tony Hill, strongly influenced by structuralist procedures, continued research into visual kinesthesis. He invented, through a poetry of tinkering, artisanal observational devices. His programmatically titled film *To See* (1982) is a visual exploration of the ambient environment. A 16 mm camera, attached to the end of a pole, faces a convex hemispherical mirror that reflects the space around, in front of and behind the camera, right up to its remote edges, inaccessible to the naked lens (Fig. 2.1).

The experience of the spatial geometry of parks and streets in different seasons is transcribed in a double-screen projection that refers to human binocular vision. The shutter flickers, added in the editing, complete this camera/eye analogy. Here, it is as if "the awareness of the surrounding environment" is synthesized into a metonymy, that of this hemispheric reflection where the edges of the visible touch each other. Rather than progressively traversing the entire visible field, is perception then a matter

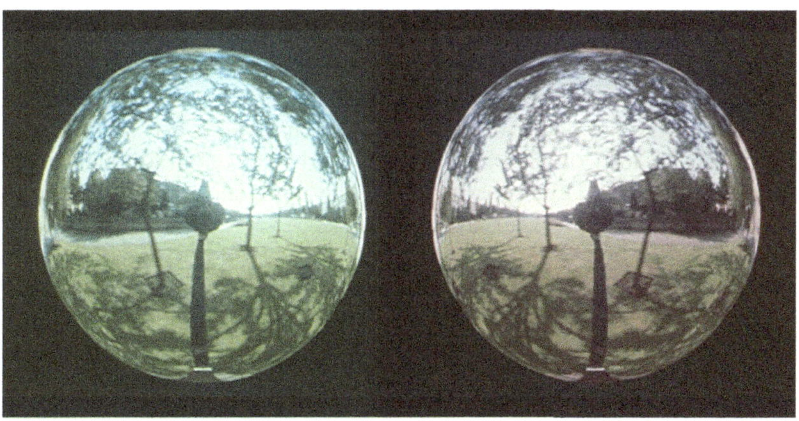

Fig. 2.1 Tony Hill, *To See* (1982) © All rights reserved by the artist/Courtesy of Light Cone

of picking up information from a space that we know pre-dates our perception?

What the specular presence of the camera indicates, in addition to a *self-reflexiveness*,[24] is that through the act of perception man is engaged in a reciprocity with the environment: a *co-perception*. Indeed, perceiving is not an abstract activity that extracts us from the world; on the contrary, it induces the perception of our own perceiving body—self-perception. Ecological philosopher David Abram says this in substance: "The body and the earth [are in] reciprocal participation".[25] The third specific movement of visual kinesthesis, ***protrusion***, is clear sensual proof of this. Perception cannot take place without a co-perception of one's feet treading the earth, of one's hands manipulating tools, of the tip of one's nose underlining the anchoring of one's head. This idea of reciprocity, where what Gibson calls egoreception (self-perception) and exteroception[26] (that of external objects) merge, is embodied by a series of filmed performances by Californian filmmaker Gary Beydler. His films *Hand Held Day* (1974) and *Mirror* (1974–75) have in common the involvement of three actants: the body and hands of the filmmaker, the manipulation of a mirror and the variation of a western American landscape. In each film, the mirror seems to concentrate the *corporeal dimension of the environment* through a frame-within-a-frame effect. The moving sensuality of the landscapes—a mountain, the crash of the sea on a beach, clouds, the sun—is amplified by the use of the interval-ometer which adds a condensation of natural time to a concentration of space. Thus, in the mirror of *Hand Held Day,* which reflects the glow of an Arizona mountain landscape (filmed from dawn to dusk), it literally appears that the filmmaker ***holds perception in his hands***. It is also the reciprocity of the idea of ***milieu*** that is invoked here: natural environment and medium. Could the mirror not be the materialization of the filmic medium as an ecological surface?

Tony Hill's film *Laws of Nature* (1997) extends this idea of co-perception. In this 35 mm film shot in a natural environment (sea, forest and meadow), human figures seem to question the foundations of their ecological perception by exploring their bodies' sensual powers. Women's hands touch bark, brush moss and rock. Further on, a man encircles the trunk of a tree with a rope, at eye level. The rope held in his hand unfolds to form a link with the tree, some 20 metres long. The man now joyfully turns around the trunk, tracing a circle, like the lead of a compass. This idea of perceptual reciprocity is concretized in this link and its winding, which we follow from the axis of the rope: a panning shot from the point of view of the trunk as it sees the man, then a circular tracking shot from the point of view of the man as he sees the trunk, in a circle that will inevitably lead us to get closer to the tree, until we touch it.

Co-operating with Nature

As we saw earlier, Malcolm Le Grice understood structuralist activity as a "process of [objects'] accessibility through acts of perception".[27] The structuralist film is thus ordered from a pre-objective basis: before being accessible to a human subject through perception, the non-human exists independently. This pre-existence of what Le Grice called "natural objects", the fact that clouds, winds, tides and currents contain within themselves *the possibilities of perceiving*, seems to be the great step forward of the English filmmaker Chris Welsby's work. His structuralist approach, which began in the early 1970s, is profoundly poetic in that it inverts the issue of perception. In Welsby's work, the manipulation of camera functions and the mechanical and perceptual processes are now entrusted to nature. The act of perception is interrogated from the outside, by nature itself and its forces, and for the filmmaker this means engineering apparatuses at the service of climatic phenomena: celestial light, tides, currents, cloud flows, Earth rotation, solstices and equinoxes, seasons. His work thus raises a double question: to what extent is the act of perception dependent on the natural environment? Consequently, isn't film, which can give voice to this exteriority of perception, a new means of non-human expression? This general trend could be qualified as *in-objectivist*: neither subjective nor objective (beyond dualism), since it is now a question of the filmmaker "co-operating" with nature.

Through the window of a water mill on the Isle of Wight, we gaze at an estuary. Successively, 35 mm then 16 mm cameras record a frame every minute (day and night) over two three-week periods, in autumn and spring. These natural periods correspond to the arrival of the equinoxes, rare moments in the year when the day is of equal length to the night, heralding a change of season. The film is shown on two screens, one per period. One minute of projected film represents one day. The movement of the tides, the luminous and chromatic variations of day and night, the curve of the sun, meteorological fluctuations (rain showers, clouds, fogs, frost and thaw), the change in vegetation; a whole expressive palette of the estuary is made visible by the time-lapse effect and the unity of the still shot. This is the film *River Yar* (1971–1972), directed by Chris Welsby in collaboration with another filmmaker from the London Film-Makers' Co-op, William Raban. This film, which was to leave its mark on the aesthetics of landscape film, opens up a new conception of ecological perception in relation to filmic processes. The climatic and astronomic value of perception is revealed through a conversion of non-human temporalities: the filmmakers find equivalences between the duration of the filmic experience and the real duration of natural cycles, days, nights and the passage of seasons. This temporal extension/condensation makes the estuary an *ecological topos* that takes nature out of pictorialism and landscape imagery. Indeed, filmed to the rhythm of the seasons, the estuary is here unveiled in the slowness of its processes and the progression of its ecological scales: we discover it as a biotope whose interacting life forms form an ecosystem, all interdependent with the climate and astronomical movements.

"What is this thing we call nature?" Chris Welsby asks in an interview with critic Michael O'Pray. He continues: "All the films are concerned with integration. The relationship between mind and nature is in a constant state of redefinition. Technology and landscape are both part of nature."[28] This borrowing from Gregory Bateson's *Ecology of Mind*[29] suggests that Welsby sees the art of filmmaking as the chance for a symbiosis between mind and nature. The filmic *dispositif* becomes for the filmmaker an *ecological organism of exchange*, where the interactions between the perceiving body, the mind, the tool and natural forces form an ecosystem. Chris Welsby was one of the few filmmakers in the early 1970s to accompany his experiments with such an ecological awareness and a practice of so-called "renewable" energies. Positioning the question of climate and natural forces at the centre of his research was not insignificant. We were at that historical moment when the first United Nations "Earth Summit", held in Stockholm in 1972, considered the impact of human activities on possible climate change. The same year saw the creation of the United Nations Environment Programme (UNEP) in Nairobi, Kenya, with a focus on water resources and renewable energies, marine mammals, desertification and forests, leading to the establishment of a legal framework for the natural environment.

By manipulating the different functions of the camera, Chris Welsby is able to uncover analogies between climatology and the perceptual singularities of the filmic medium. He proposes a series of *wind-oriented observational apparatuses*: *Wind Vane* (1972) and the triptych *Windmill*,

I, II, III (1972–1974). In *Wind Vane*, the panning movement is activated by the wind itself, whose variations act on a vane blade attached to the camera. In the *Windmill* series, also shot in a London park, it is no longer the panning movement that is a meteorological phenomenon, but the functions of the shutter: the blades of a wind turbine covered in melanex, a mirror adhesive coating, face the lens of a 16 mm camera mounted on a tripod. The variation in the speed of the blades depending on the wind (analogy for the shutter) leads to different states of perceptive consciousness in the viewer, in their apprehension of depth, surfaces and natural space. The line of trees in the background, the random movement of passers-by behind and in front of the camera, combine with the variable speed of the blades and the filmmaker's reflection captured in the foreground (Fig. 2.2).

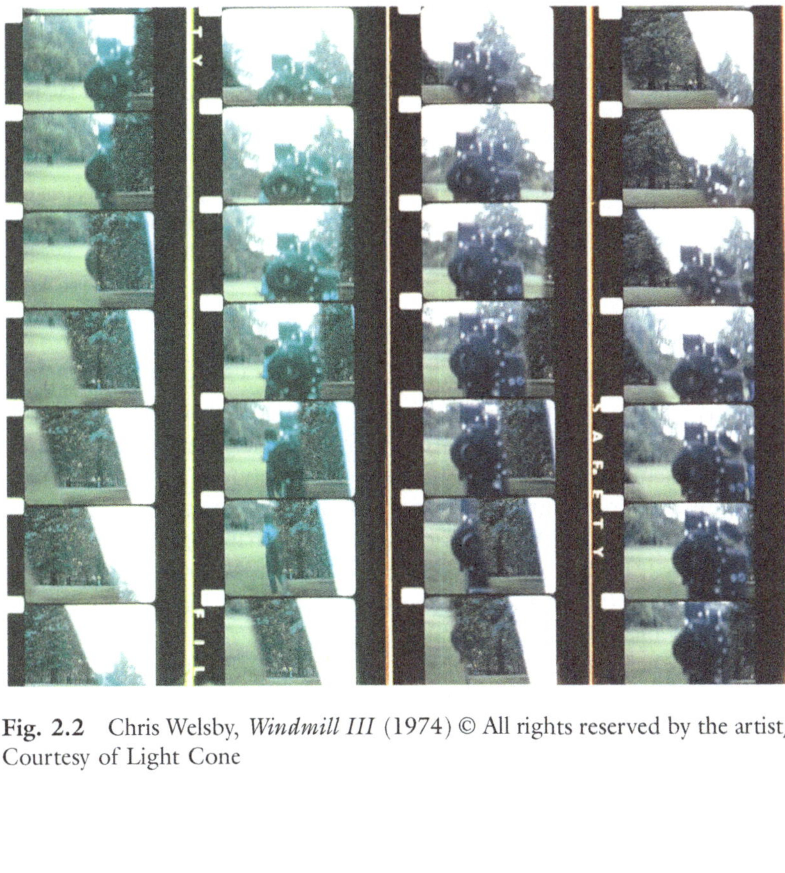

Fig. 2.2 Chris Welsby, *Windmill III* (1974) © All rights reserved by the artist/ Courtesy of Light Cone

Wind Vane and *Windmill* entrust the act of perception (here the scanning, the perception of movement or questions of depth and surface) to meteorological fluctuations. As in direct perception, perceiving becomes an experience of adapting to the changing data of the environment. The film is no longer an objective framework for observing the agency of natural phenomena. It is the agency of nature that *in-objectively* orients the form and content of the film.

Gibson develops a radical hypothesis that illuminates his ecological approach to perception: the theory of *affordances*. *Affordances* are "properties of the real environment as directly perceived by an agent in a context of practical action".[30] By affordances, Gibson means that the values and meanings of things are external to the perceiving subject. They are *what is offered* to him in view of a practical action. Perceiving is therefore no longer an internal activity dictated by a sensorimotor circuit, perceiving is acting according to the benefits or harms *offered* by environments, surfaces, substances, objects, places or other animals. Similarly, to perceive is to act according to the singularities of one's ecological niche, one's particular habitat within an ecosystem. The cinema of Chris Welsby seems to orient the question of perception towards the same attention as what is offered by nature: its flows, its surfaces, its ecological niches, its environments. It opens up the idea of the film as a capture of properties external to the perceptive phenomenon, what is offered to perception: the affordances.

The Speech of the World

An ecopoetic threshold seems to have been crossed, from the objectivism of Michael Snow to the in-objectivism of Chris Welsby; the filmmaker now goes so far as to entrust nature with the indexicality of the images. The film resembles a new signifying system, composed of "living" signs: wind signs, equinox signs, river signs. In this respect, the act of filming becomes a translation game, translating a non-human expressivity: a *speech of the world*.[31] This indexicality of the non-human accompanies the post-structuralist practice of the Californian filmmaker and composer Robert Schaller. In his work, the mountain can be seen as a secret diamond. Robert Schaller's father was one of the great American mountaineers involved in the CIA's secret mission to spy on the Chinese military programme. In 1965, after weathering a mountain storm, the team finally climbed Nanda Devi, India's highest Himalayan peak (7816 metres), carrying several prototypes of observation, listening and communication apparatuses. His son inherited his passion for the mountains and his experimental ingenuity. In 1998, he built a tool that would shape his practice, a handmade *pinhole camera*: a light-tight black box made from 16 mm film canisters, in which he could manually pass a 16 mm film in front of a tiny needle hole (0.06 mm). Each time a frame passes through, an image is imprinted on the 16 mm negative film (thanks to an ingenious system of adjustment-window and butterfly nuts), according to an exposure time that he can vary. The apparatus implies a certain amount of time between each frame, and a spatial decomposition effect when screened.

With the use of the pinhole camera, Robert Schaller's practice increasingly focuses on the question of a basic ecological perception. The filmmaker has produced variations on the mountain theme: *Peak* (2008), *Mountain Home* (2010) and *In the Shadow of Marcus Mountain* (2011). Moving beyond the framework of his first films, where the musical score was embodied in the rhythms of the natural world, Schaller gives way to a more autonomous non-human expression. Through the pinhole camera and the immanent materiality that he entrusts to the 16 mm film, the mountain seems to be able to speak for itself. "Our eyes see constantly, but what do we actually notice? That vision is excessive, wasteful, even; in paring down, it becomes both more spare and more concentrated."[32] The pinhole technique no longer relies solely on the eyes to perceive. Schaller takes the device in hand, and it is his whole body, in an intertwined relationship with the natural environment, that sets the act of filming in motion. The images, inflamed, *im*-mediate, fragmentary, concentric, threshold images, reveal a bodily experience of the *very act of perceiving*. A "mountain speech" is translated here, captured in its precariousness. The silhouettes of trees, ridges and their contours, the relationships between surfaces and contrasts, the grainy depth of the sky, all emerge and affect the black-and-white film, leaving something like a volcanic trace, a blaze. The living signs of the mountain manifest themselves, addressing their own "response" to Tess Takahashi's inquiry: "(…) does celluloid film allow nature to "speak" more directly?"[33](Fig. 2.3).

Listening to this mountain speech through a pinhole is also to reread Gibson's ecological theory and his central concept of the ambient optical array. Perceiving, as we said earlier, is picking up information contained in the natural world. According to Gibson, this picking up is made possible by an optical phenomenon unique to animals: the ability to grasp *invariant* aspects in the ambient optical array. Ambient light at a point of observation is structured in relation to an observer (stationary

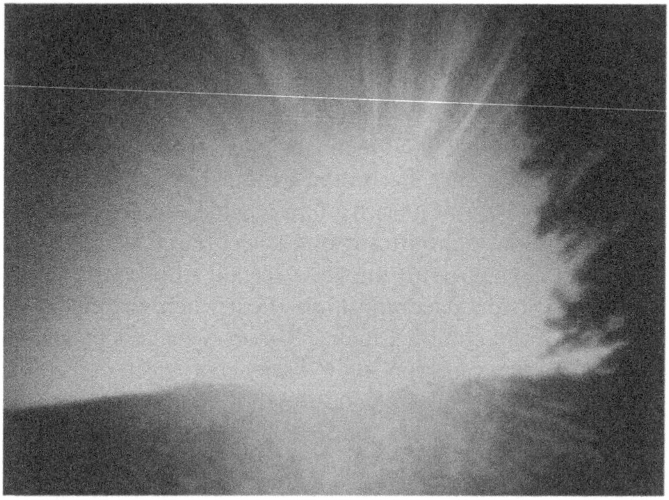

Fig. 2.3 Robert Schaller, *In the Shadow of Marcus Mountain* (2011) © All rights reserved by the artist/Courtesy of Light Cone

or mobile) thanks to the presence of surfaces that emit and reflect light. These surfaces and their arrangement are necessary to extract invariants. Let's listen to Gibson again: "The components of the array are the *visual angles* from the mountains, canyons, trees and leaves, [...] They are *intercept angles*, as we shall see", then: "All these optical components of the array, whatever their size, become vanishingly small at the margin between earth and sky, the horizon".[34] These ecological conditions that inaugurate the act of perception—surface reflectance, array, intercept angles, horizon line—seem to find a poetic equivalence in Robert Schaller's films. *In the Shadow of Marcus Mountain* questions the conditions under which these *invariants* appear, through an exploratory process. Starting from the observation of a familiar place, a peak in the Rocky Mountains near his home in Colorado, Schaller recomposes the memory of things perceived: these pine trees, these dry branches, this clearing, this mountain slope, this rocky diagonal, this sensuous ridge drawn by the horizon; a variety of instantaneous viewing angles, alternating with black frames, proposes a perceptual cycle that goes from sleep to full wakefulness. *In the Shadow of Marcus Mountain* ultimately synthesizes what seems to be a paradox of visual perception: isn't the most concentrated act of perception, in the end, *letting the world speak*?

Towards a Cosmopolitics of Representation

In the twenty-first century, photochemical practices are vigorously pursuing the path opened by structuralist filmmakers. Both the ecological dimension of visual perception and that of the medium as a material surface of exchange, as well as a renewal of observational devices, are reincarnated in the works of filmmakers like Philip Hoffman, Emmanuel Lefrant, Emily Richardson, Inger Lise Hansen or Daniel Steegman Mangrané.[35] But looking back here at the historical context of the very first structuralist films allowed us to observe the original emergence of an ecological consciousness in the practice of experimental filmmakers. This consciousness, which went hand in hand with an experience of visual perception in the natural environment, emerged on a much less ecologically affected Earth. From the pioneers of the 1960s to today, media coverage of climate and environmental issues has continued to grow at an exponential rate similar to that of the damage it strives to inform us of: carbon emissions from human activities, the retreat of the cryosphere and rising sea levels, increasing fires and floods, industrial pollution of environments, desertification due to intensive agriculture and global warming, deforestation, the dangers of nuclear power, accelerated disappearance of wild species, etc. The Sixth Assessment Report of the IPCC, published in August 2021, announced an acceleration in global warming, revising the estimates of the 1990s (themselves based on an increase in temperature since the pre-industrial era). Climate experts are announcing that extreme intensity climate episodes will occur in about ten years' time, meaning that the 1.5 °C maximum increase, which states have pledged not to exceed, will almost certainly be reached by 2035. "What we perceive and how we perceive"[36] is like an appeal launched in this troubled geo-climatic period, the Anthropocene, where image networks have become hypermediatic. The question posed by T. J Demos: "How does the Anthropocene enter into visuality, and what are its politics of representation?",[37] announces a programme of visual resistance with numerous ramifications, involving relations of domination with the non-human but also and above all the colonization of other human ontologies. Re-enacting the act of ecological perception, by expressing the reciprocity between humans and their natural environment, would be one such form of resistance. Defended by filmmakers who are considered marginal and who entrust the non-human with the means of its own expression, the act of ecological perception could open the way to a form of *cosmopolitics of representation*.[38] Through the moving image, the fruit of an exploratory process, we glimpse a ray of light: after having been "humans in nature", we can hope to become "earthlings among earthlings".[39]

Notes

1. Roger Munier, *Contre l'image*, Paris, Gallimard, 1963, p. 99. Unless otherwise noted, all translations from the original French are by the author and translator.
2. James Jerome Gibson, *The Ecological Approach to Visual Perception*, New York, Taylor & Francis Group, 1986 (1979).
3. Hermann Von Helmholtz, *Treatise on Physiological Optics* [1867], trans. James P. C. Southall, New York, Dover Publications, 1961.
4. See Marion Luyat, Tony Regia-Corte, "Les affordances: de James Jerome Gibson aux formalisations récentes du concept", *Année psychologique*, Vol. 109, 2009, pp. 297–332, available online at: https://www.cairn.info/revue-l-annee-psychologique1-2009-2-page-297.htm [consulted online on 10 June 2021].
5. See the discussion of J. A. Fodor and Z. W. Pylyshyn, "How direct is visual perception? Some reflections on Gibson's 'Ecological Approach'", *Cognition*, no. 9, 1981, pp. 139–196.
6. See the synthesis of J. J. Gibson's ecological approach to visual perception made by the ecological anthropologist, Tim Ingold, "Point, Line, Counterpoint. From Environment to Fluid Space", in *Being Alive. Essays on Movement, Knowledge and Description*, Oxon, Routledge, 2011, pp. 76–88.
7. James Jerome Gibson, *The Ecological Approach to Visual Perception, op. cit.*, see especially Chapter 16: "Motion Pictures and Visual Awareness", pp. 292–302. This idea of a "progressive picture" is thought in distinction from the "arrested picture" of painting and photography. As a depiction activity, cinema presents a flow of varied optical events, disturbances of structures, textures, an optical array, analogous to those of human perception in a natural environment.
8. Viva Paci locates this emergence of a "machinist" conception of cinematographic perception at the heart of the "*Cinéma pur*"

of the 1920s–1930s, rereading critics like Luigi Pirandello or filmmaker-poets such as Jean Epstein, Germaine Dulac, Louis Delluc: "Chapitre 2 - La nature du cinéma: entre photogénie et attraction", in Viva Paci, *La machine à voir, à propos de cinéma, attraction, exhibition*, Villeneuve-d'Ascq, Septentrion University Press, 2012.

9. See here Clélia Zernik, *L'œil et l'objectif: La psychologie de la perception à l'épreuve du style cinématographique*, Paris, Vrin, 2012. In our present study, returning to the shooting stage questions the naturalness of perception and allows us to move beyond the experimental limits of Gestalt psychologists. These studied the relationships between human perception and cinematographic perception in the applied context of the movie theatre, an analogy for the "laboratory".
10. Kim Knowles, *Experimental Film and Photochemical Practices*, London, Palgrave MacMillan, 2020, p. 75.
11. Bruno Latour, *Facing Gaia: Eight Lectures on the New Climatic Regime* [2015], trans. Catherine Porter, Cambridge, Polity Press, 2017. The expression "New Climatic Regime" allows us to observe contemporary geopolitical issues through the prism of climate disruption.
12. Scott MacDonald, "Toward an Eco-Cinema", *ISLE: Interdisciplinary Studies in Literature and Environment,* Vol. 11, No. 2, Summer 2004, p. 109.
13. Ibid.
14. The adjective "structural", in addition to designating a major current of the avant-garde of the 1960s–1980s, refers to the theoretical debate animated (at the turn of the 1970s) by critics such as Paul Adams Sitney, George Maciunas, Peter Gidal and Malcom Le Grice, positioning themselves in rupture with the formalist current. The characteristics of structural cinema, retained by P. A. Sitney, (but discussed in later texts) are: the fixed camera position (fixed frame from the viewer's perspective), the flicker effect, loop printing and rephotography off the screen.
15. I am here using the term used by Malcolm Le Grice, in Chapter 10 "Current developments", in Malcolm Le Grice, *Abstract Film and Beyond*, London, Studio Vista, 1977, p. 124.
16. The London Film-Makers' Co-op is one of the first experimental and independent film cooperatives, founded in 1966 by Stephen

Dwoskin and Bob Cobbing. Based on the model of Jonas Mekas's *Film-Makers' Cooperative* in New York, this collective laboratory innovates by making all stages of the film practice autonomous: production, distribution and exhibition. The filmmakers Peter Gidal and Malcolm Le Grice are among the notable figures of this British avant-garde, contributing to the theorization of a so-called "structural/materialist" cinema and an "expanded cinema". Several aesthetic avant-gardes flourished there in an exploration of the materiality of the body and the apparatus, of the self-reflexivity of the process, of the listening and viewing situation, of new modes of exhibition and projection: a subversive cinema by John Smith, a landscape cinema by Chris Welsby and William Raban, a performance and feminist cinema by Lis Rhodes, Gill Eatherley, Marilyn Halford and Annabel Nicolson.

See the editorial of the retrospective organized by Mark Webber from 3 to 28 May 2002 at London's Tate Modern Gallery: "SHOOT SHOOT SHOOT, the first decade of the London Film-Makers' Co-operative & British avant-garde film 1966–76". Available online at: https://monoskop.org/File:Shoot_Shoot_Shoot_The_First_Decade_of_the_London_Film-Makers_Co-operative_and_British_Avant-Garde_Film_1966-1976.pdf [consulted online on 2 May 2021].

17. Deke Dusinberre, "St. George in the Forest: The English Avant-garde", London, *Afterimage*, no. 6, Summer, 1976, pp. 11–13.
18. Ibid., "The landscape is thus an integral factor in determining the shape of the film" says Dusinberre.
19. Malcolm Le Grice, "Kurt Kren's films", in Peter Gidal, *Structural Film Anthology*, London, British Film Institute, 1978, pp. 56–63.
20. Ibid., p. 61.
21. Malcolm Le Grice, *Abstract Film and Beyond, op. cit.*, p. 124.
22. J. J. Gibson, *The Ecological Approach to Visual Perception, op. cit.*, pp. 112–114.
23. Michael Snow, quoted in Charlotte Townsend, "Converging on *La Région Centrale*: Michael Snow in Conversation with Charlotte Townsend" (1971), in Michael Snow, *The Collected Writings of Michael Snow*, Ontario, Wilfrid Laurier University Press, 1994, p. 57.
24. The term "self-reflexiveness" is borrowed here from Peter Gidal. It refers to this tendency of the structural film to present its own

making, to put into perspective the tools of its own analysis during the projection, and to use material operations to present "consciousness to the self". Peter Gidal, "Theory and Definition of Structural/Materialist Film", in Peter Gidal (ed.), *Structural Film Anthology*, London, British Film Institute, 1978, p. 10.

25. David Abram, *The Spell of the Sensuous, Perception and Language in a More-Than-Human World,* New York, Vintage Books, 1996, p. 81.
26. J. J. Gibson, *The Ecological Approach to Visual Perception, op. cit.,* in which the concepts of egoreception and exteroception are detailed in Chapter 7: "The Optical Information for Self-Perception", pp. 111–126.
27. M. Le Grice, "Kurt Kren's films", *art. cit.,* p. 61.
28. Chris Welsby, "Interview with Chris Welsby, Michael O'Pray and William Raban", *Undercut 7/8,* Special Issue "Landscape in Film, Photography, and Video", London, London Film-Makers' Co-op, Spring 1983.
29. Gregory Bateson, *Steps to an Ecology of Mind,* San Francisco, Chandler Publishing Company, 1972.
30. T. Ingold, "Culture and the Perception of the Environment", in Elisabeth Croll, and David Parkin (eds.), *Bush Base: Forest Farm. Culture, Environment and Development,* London, Routledge, 1992, p. 46.
31. The expression "Speech of the world" is borrowed from the philosopher-poet Roger Munier and his ecopoetic ontology of the photographic image, in *Contre l'image, op. cit.*, and *Le Chant Second*, Paris, Deyrolle, 1991.
32. Robert Schaller, descriptive note of the film *In the Shadow of Marcus Mountain*: https://www.robertschaller.org/film/ [Consulted on 10 May 2021].
33. Tess Takahashi, "After the Death of Film: Writing the Natural World in the Digital Age", *Visible Language*, Vol. 42, no. 1, 2008, p. 48.
34. J. J. Gibson, *The Ecological Approach to Visual Perception, op. cit.*, p. 68.
35. See *Redshift* (2001), *Aspect* (2004) by Emily Richardson; *Proximity* (2006) by Inger Lise Hansen, *16 mm* by Daniel Steegman Mangrané (2009–11); *I don't think I can see an Island* (2016) by Emmanuel Lefrant; *Vulture* (2019) by Philip Hoffman.

36. K. Knowles, *Experimental Film and Photochemical Practices*, *op. cit.*, p. 75.
37. T. J. Demos, *Against the Anthropocene: Visual Culture and Environment Today*, Sternberg Press, London, 2017, p. 10.
38. The use of the term "cosmopolitics" is part of a debate on the philosophy of nature initiated by Isabelle Stengers and taken up by thinkers such as Bruno Latour or Pierre Montebello. The Kantian universalism of the term 'cosmopolitical' is superseded by an openness to the 'cosmos' as a political expanse bringing together the human (considered in its various ontologies) and the non-human. In these environmentally troubled times, it has become even more apparent that Western Human is not the only political agent on Earth. A "cosmopolitics of representation" would thus be a critical and ethico-political approach to the filmic image, extended to all political agents presented therein. We can hope that, through its own means, cinema contributes to spreading a double resistance: to the domination of nature by humans, and to that of humans by humans, fighting against the representative propaganda of conventional media. See Jacques Lolive and Olivier Soubeyran, *L'émergence des cosmopolitiques*, Paris, La Découverte, 2007.
39. Bruno Latour, *Down to Earth. Politics in the New Climatic Regime* [2017], trans. Catherine Porter, Cambridge, Polity Press, 2018, p. 86.

CHAPTER 3

Eco-Cinematic Long Shots

Scott MacDonald

1.

When I chose the term "eco-cinema" for my essay, "Toward an Eco-cinema", published in the Summer 2004 issue of *ISLE* (*Interdisciplinary Studies in Literature and Environment*), I don't believe I meant to invent a term.[1] "Eco" seemed everywhere in the early 2000s; and it is in the nature of language to respond to the interweaving of cultural developments. "Eco-cinema" seemed a good way of referencing the intersection of the continuing evolution of cinema with a cultures-wide recognition that many ecosystems on our planet, and the planet itself, are in a process of transition, if not entropy. That "eco-cinema" has become a frequently used term was perhaps inevitable, as moving-image artists, including artists identified with the history of avant-garde and experimental cinema, have increasingly turned their attention to the realities of ecological transformation.

Of course, moving-image artists of many kinds have produced what one might call eco-cinema. No one could watch Jeff Orlowski's *Chasing*

S. MacDonald (✉)
Hamilton College, Clinton, NY, USA
e-mail: smacdona@hamilton.edu

E. Della Noce and L. Murari (eds.), *Expanded Nature*, Experimental Film and Artists' Moving Image,
https://doi.org/10.1007/978-3-031-70729-2_3

Ice (2012) and not have their understanding of the impacts of climate change on global ecology reconfirmed. And after seeing the panorama of places where human exploitation of global resources has resulted in ecological ruin, in Nikolaus Geyrhalter's *Homo Sapiens* (2016), one cannot help but consider, as some science fiction writers have long imagined, that humanity is in the process of mass suicide. Geyrhalter makes this comment: "*Homo Sapiens* is very much about the strength of nature and about the loss of humanity, which is maybe not as big a loss as we usually imagine".[2] Even the long history of the commercial nature film that has been so popular on television has begun to move away from an earlier tendency to see "nature" as a beautiful and pristine world entirely apart from humanity. The second episode, "Frozen Worlds", of Richard Attenborough's recent *Our Planet* (2020) ends with a sequence of walrus, forced by the warming of the Arctic to squeeze themselves onto precarious high ground, falling to their deaths—as horrific an image of natural destruction as one can imagine.[3]

Within the history of what is usually called avant-garde or experimental cinema, the most interesting "eco"-oriented work, or at least the work I have been most aware of and engaged with, has been of a radically different order. A bit of background may be useful. As the global awareness of the relationship between consumer culture and environmental damage began to grow during the second half of the twentieth century, some independent filmmakers worked to confront the mass-media forms that were proliferating in the service of material acquisition and consumption. During the 1950s and 1960s, the increasing propaganda for conspicuous consumption was evident in a variety of ways, most obviously perhaps in the design of the commercial television hour, constructed so that comedy and drama series would deliver viewers to advertising montages. These increasingly heavily edited commercials were generally resonant of the Soviet montage editing of Sergei Eisenstein; indeed, their approach of overloading viewers with dozens of images per minute was itself a psychic modeling of conspicuous consumption. By the end of the twentieth century, anyone growing into adulthood within a modern, commercialized society would have seen hundreds of thousands of television commercials.

The new millennium has continued this tendency, though the forms within which we receive commercial messages have proliferated. It is not unusual to see young people engaged with their computers, while texting on their cell phones in a room where a television is on or a video game is being played—immersed within the contemporary commercialized audio/visual environment. And the number of commercial sites for viewing, sending/receiving messages, and doing mini-performances continue to proliferate across the spectrum of media devices and systems.

Beginning in the late 1960s and especially during the early 1970s, a number of independent filmmakers were becoming committed to forms of cinema that could work, at least implicitly, against the onslaught of commercial advertising and its assumption that a successful life requires that one have more and better things to consume, more and more of the time. Filmmakers working to imagine moving-image experiences that were not devoted to acquisition and consumption often turned to the earliest cinema and pre-cinematic forms (the Lumière Brothers films and Eadweard Muybridge's motion photographs, in particular) for inspiration.[4] The result was what became a genre of independent filmmaking devoted to the use of the single, extended shot in order to redirect the experience of moving-image media toward meditative and contemplative motion study of the filmmaker's surround. Larry Gottheim's *Fog Line* (1970) has become a canonical instance.

Fog Line was not the first single-shot film to work against the mass-media dedication to sensory overload. Bruce Baillie's *All My Life* and *Still Life* (both 1966) predate *Fog Line*, as do Robert Nelson's *The Awful Backlash* (1967) and Gottheim's own *Blues* (1969). And there are Warhol's single-shot *Screen Tests* (1964–66) and other long-take films, which had considerable influence on all of these filmmakers—though Warhol's films can seem as dedicated to unbridled consumption as commercial media. *Fog Line* remains distinctive because of its elegant way of meditating on nature/culture in a manner that foregrounds the need for a more perceptive and patient kind of viewing, devoted not to the consumption of image and action, but rather to the transformation of our way seeing the world around us and providing an instigation for our living more thoughtfully within it.

The eleven minutes of *Fog Line*, basically the length of a 400-foot roll of 16 mm film, with front and rear flares eliminated, begin with what appears to be an empty white screen, though several vague shapes are visible. Gradually we realize that we are seeing a small dell covered in fog that is slowly clearing, to reveal several trees and a green field. The image is graphically divided into horizontal thirds by two power or phone lines (just above the top line on the left is what may appear to be the sun barely visible behind the fog, but is in fact a smudge on the lens). Particularly alert viewers may see that, about half way through the film, two horses graze slowly across the bottom third of the image from right to left. Recognizable once one compares the size of the trees in the middle of the composition to the size of the horses in the foreground is the fact that the film was shot with a telephoto lens.

What many viewers tend to identify as a natural scene is, in fact, an implicitly self-reflexive emblem of the intersection of natural process (the clearing of the fog, the movement of the horses) and human technology. The power/phone lines, echoing the rectangular frame of the 16 mm image, suggest a minimal Cartesian grid. And of course, this being a north-eastern landscape, the pasture in which the horses graze must be fenced, just as the film frame "fences in" the tiny events of the film.

For Gottheim, the gradual transformation of the *Fog Line* image and the subtle development of our sense of what we're seeing is both visual experience and metaphor:

> For me, making a film is like opening up a veil. I'm seeing something or experiencing something, and I'm not sure what it is. Actually, during the period when I was making *Blues* and *Fog Line*, I think that I was very happy to talk about mysticism or mystical things.... These days I tend to talk much more about energy.... Certainly, energy is a very real element in life, not just the electricity that powers light bulbs but the life energy in people that has to do with their relationships to each other and their relationships to nature.[5]

Until he moved to the Binghamton area to teach at Binghamton University (then the State University of New York at Binghamton), Gottheim had been an urbanite. To some degree, *Fog Line* and his other early, single-shot films—*Blues*, *Corn* (1970), *Doorway* (1971), *Thought* (1971), and *Harmonica* 1971)—as well as *Barn Rushes* (1971) and *Horizons* (1973), can be understood as an ongoing perceptual and spiritual exploration of the pastoral landscape that surrounded him, just beyond the small urban center of Binghamton—and, more generally, of the implications (perceptual, practical, historical, spiritual) of landscape for humans and vice versa.[6]

2.

The 1970s and 1980s saw a number of filmmakers exploring the extended shot—usually, series of extended shots—as a means of retraining cinematic perception away from both the hard sell of commercial advertisements and the demands of narrative, where sequential editing delivers viewers to the sense of a story consumed and digested. Early on, Peter Hutton and James Benning, in somewhat different ways, were dedicated to an expanded sense of the environments they portrayed and to a new kind of viewership.

From the beginning of his work in the early 1970s, Peter Hutton abjured conventional forms of film and television excitement:

> Most people go to films to get some kind of overwhelming experience, whether it's like an amusement park ride or an ideological, informational hit that gives you insight into an issue or an idea....
> The experience of my films is... about taking the time just to sit down and look at things, which I don't think is a very Western preoccupation. A lot of influences on me when I was younger were more Eastern. They suggested a contemplative way of looking—whether at painting, sculpture, architecture, or just a landscape—where the more time you spend actually looking at things, the more they reveal themselves in ways that you don't expect.[7]

The primary focus of Hutton's films is cityscape and landscape. In his series of New York City films—*New York Near Sleep for Saskia* (1972) and *New York Portrait I* (1977), *II* (1980), and *III* (1990)—Hutton's focus is not the excitements of urban life or a celebration of the modern metropolis, but quiet moments of subtle change, emphasized by his commitment, at least until his final years, to black-and-white imaging and, throughout his career, to silence. His landscape films, mostly focusing on the Hudson River valley, follow suit; they offer a panorama of individual, contemplative, silent, black-and-white shots, often separated from one another by moments of darkness. *Landscape (for Manon)* (1987) is typical (Fig. 3.1).

Landscape (for Manon) begins quietly, with shots that last 20 to 30 seconds, but where a narrative film might speed up to create more action and excitement, Hutton slows his film down, to shots that last nearly a minute, enticing viewers toward a more thoughtful awareness of the subtleties of the living world revealed in the film. His dedication of the

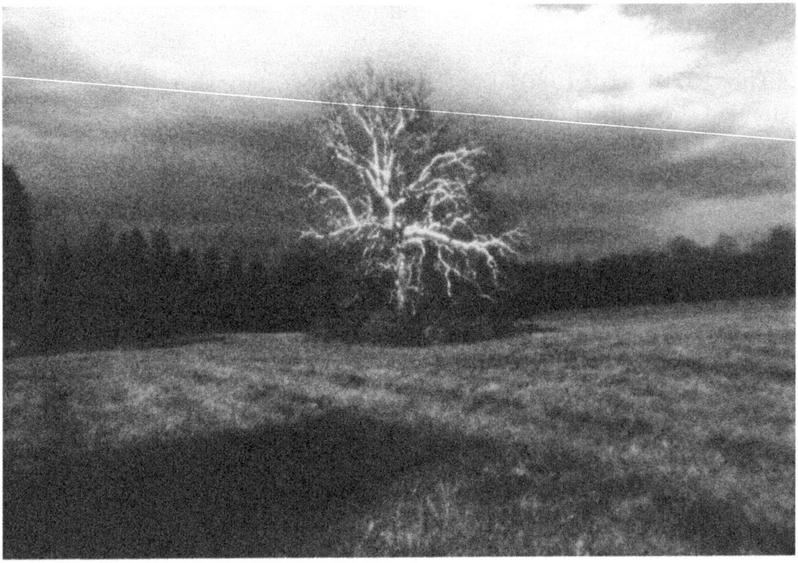

Fig. 3.1 Peter Hutton, *Landscape (For Manon)* (1987)

film to his young daughter, Manon, suggests that viewers can reclaim something of childlike wonder, an earlier, less exploitive way of seeing, implicitly modeled by the film. In his final Hudson River film, *Time and Tide* (2000), Hutton returned to color in order to help visualize the way the river valley had been, and to some extent still was, the location of various industrial adventures. Indeed, during the time when the film was shot, efforts were afoot to clean up the results of a century of industrial exploitation of the river, as well as to resist further exploitation. In general, Hutton's landscape films—*In Titan's Goblet* (1991), *Study of a River* (1996), his Islandic film *Skagafjördur* (2004), and *Three landscapes* (2013)—evoke Thomas Cole, Frederick Church, and other painters identified with the American Hudson River School, the first art movement in North America instigated by concern for the natural environment and the dangers of industrialization.

Like Hutton, Benning was, from early on, fascinated by landscape and cityscape. In his *8½ × 11* (1974), *11 ×14* (1976), and *One Way Boogie Woogie* (1977), he deployed series of extended shots to represent and contemplate urban and rural mid-America—though unlike Hutton's films, Benning's have always included sound and are nearly always shot in color. Also, unlike Hutton, who resolutely resisted any suggestion of narrative, in these early films Benning begins by creating what seem to be characters and potential narrative moments that do not evolve into story, focusing instead on moments of presence where humans function as elements of cityscape and landscape. For *One Way Boogie Woogie* Benning created a formalist design of 60 one-minute shots that together offer a sound-image panorama of what was becoming a rust-belt industrial area of Milwaukee not far from where he had grown up. The humans we see in the film are merely elements within moments of sound-image presence—at times self-reflexive elements, as for example, when a man walks past Benning's camera, then walks past again, clearly having walked behind the camera to return to the image. In all these films, Benning, like Hutton, is asking for a new kind of viewer patience and engagement, a new kind of looking and listening.[8]

Benning's use of extended shots would evolve over the decades in dozens of films. Each film in his California Trilogy—*El Valley Centro* (1999), *Los* (2000), and *SOGOBI* (the Shoshone word for "Earth",[9] 2001)—focusing respectively on the agriculture of the Central Valley, urban Los Angeles, and what was left of what might be called wilderness, is made up of 35 rigorously composed 2½-minute shots that cinematically map the state. But after he saw Sharon Lockhart's *Goshogaoka* (1997), which was made up of 10-minute carefully composed choreographic shots of a Japanese basketball team, he was impressed by how Lockhart had "radicalized structure, pushing duration to a new aesthetic level. After I saw *Goshogaoka*, my own work became more radical".[10] For his 2004, more-or-less diptych, *13 Lakes* and *TEN SKIES*, Benning used the 10-minute shot as his basic module, and once he left 16 mm filmmaking for digital production after *RR* (2007), he moved toward even more demanding durations: *Ruhr* (2009), his first digital feature, begins with six 10-minute shots in and around Duisburg, Germany, and concludes with a one-hour shot of a cooling tower, used in the making of coke—at the time a new, relatively environmentally friendly method of processing coal.

Both Lockhart and Benning would continue to expand on their fascination with durational observational cinema. After *Goshogaoka*, Lockhart's own work became more temporally adventurous: *Teatro Amazonas* (1999) was a 40-minute, 35 mm shot of an audience in the opera house in Manaus, Brazil, made famous by Werner Herzog's *Fitzcarraldo* (1982). In *NŌ* (2003) Lockhart documents what seems to be a single, 35-minute shot of two farmers mulching a field (there is one nearly invisible cut). She would return to the 10-minute shot as a structural module in her feature *Pine Flat* (2006), part of a photographic/cinematic project on young people in rural, mountainous California. And in *Double Tide* (2009), Lockhart documented Jen Casad clamming in Seal Cove, Maine, in what appear to be two continuous 45-minute shots (there are several invisible edits), one made in the morning, the other in the evening on that rare day when two tides make two sessions of clamming possible.[11] *Double Tide*, shot in 16 mm and transferred to high-definition video, has been presented both as a feature film and as a gallery installation where the two "shots" are exhibited simultaneously on two screens. Lockhart's respect and nostalgia for Casad's hands-on labor is obvious, as is her interest in immersing viewers within the image-scape and soundscape of Seal Cove (Fig. 3.2).

Like Hutton's Hudson River films, *Double Tide* evokes the Hudson River painters.

Benning's *BNSF* (the title references one of the largest railway networks in North America) is, to date, his longest "single-shot" work (as in *Double Tide* there are invisible "edits", made when he changed S×S memory cards in his digital camera). The tripod-mounted camera reveals a Mojave Desert landscape near Amboy, California over a period of three hours and twenty minutes, during which the space is continuously transformed by the subtle movement of wind and clouds (for example, particular distant mountains are sometimes visible, sometimes not). Every few minutes a train is heard approaching, then is seen/heard moving noisily through the image, either from left to right or right to left, after which the noise of the train gradually fades and we return to the more subtle changes in light and sound. In a sense, *BNSF* is the macroscopic version of the more microscopic *Fog Line*, at least in the ways in which natural process and industrial technology intersect in the two films. In *Fog Line* a relatively small, pastoral space-in-motion is enclosed and divided by the lines that cross and surround the image. In *BNSF* Benning films what seems, at first, an immense, nearly wilderness space, but when the first train approaches, then moves through the image pulling its immense load along the rail line, we realize that even this expansive, seemingly empty desert landscape is part of the network of modern industrial America.

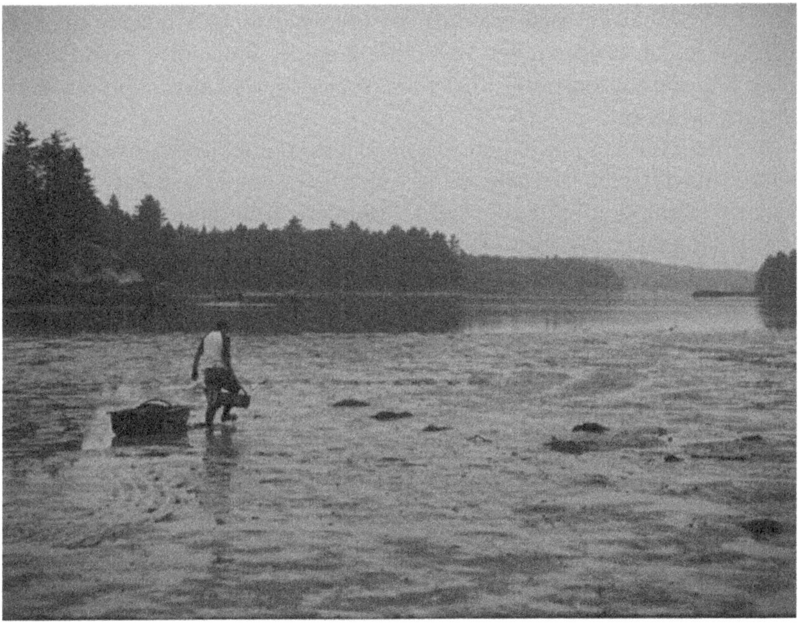

Fig. 3.2 Sharon Lockhart, *Double Tide* (2009) © All rights reserved by the artist/Courtesy of Light Cone

3.

Contributing to the more recent evolution of the eco-cinematic long shot is a modern response to what had, by the 1970s, come to be called "ethnographic film". During the 1950s and 1960s, the ethnographic interest in "pre-modern" societies living in close harmony with the natural environment instigated books and films by Lorna and John Marshall, Robert Gardner, Timothy Asch, and Napoleon Gagnon, among others. The films were instances of "salvage ethnography", a way of evoking what the filmmakers interpreted as fundamental dimensions of the lives of peoples living within natural worlds unchanged for millennia, about to face a radical cultural transformation. In time, a younger generation (David and Judith MacDougall, Leonard Kamerling, Sarah Elder) questioned whether detached analysis (in films or books) by cinematic visitors to "pre-modern" societies could do anything but confirm what this new generation saw as romantic preconceptions. Indeed, the very presence of anthropologists and ethnographic filmmakers in these far-flung regions demonstrated that these societies were already in a process of transformation.

By the 2000s a third generation of ethnographically informed and committed filmmakers came to be identified with Harvard University's Sensory Ethnography Lab (SEL), founded and directed by scholar-turned-filmmaker Lucien Castaing-Taylor. The SEL filmmakers may have admired the courage and stamina of earlier ethnographic filmmakers and some of their films, but they turned away not only from the assumption that a film or several films can provide something like a comprehensive sense of an unfamiliar society, but also that ethnographic filmmaking should focus on "pre-modern" Others. The SEL filmmakers understood their filmmaking as the creation of immersive cinematic experiences of modern life, and particularly the ways in which modern life in all its complexity was transforming and being transformed by the natural environment.

Several of the most accomplished SEL filmmakers shared with Gottheim, Hutton, Lockhart, and Benning a commitment to the extended shot, and with Benning and Lockhart, an interest in representing landscape and cityscape with both image and sound. Indeed, the SEL filmmakers expanded the importance of sound, assuming that an "immersive" cinematic experience was best achieved by making sound at least as crucial to the cinematic experience as long-take contemplative shots. SEL films typically begin with sound, immersing viewers in a soundscape before imagery is presented. *Sweetgrass* (2009) by Castaing-Taylor and Ilisa Barbash, *Leviathan* (2012) by Castaing-Taylor and Véréna Paravel, *Songhua* (2007) by J. P. Sniadecki, and *Manakamana* (2013) by Stephanie Spray and Pacho Velez are useful examples of the SEL approach.

Sweetgrass was the first film identified with the SEL to garner considerable attention, though it was shot when Castaing-Taylor and Barbash were still living and working in Boulder, Colorado, and not completed until they had relocated to Cambridge. Implicitly a work of salvage ethnography, *Sweetgrass* is a cine-document of the final instance of a practice that had been going on for a century in the area of Big Timber, Montana: the epic annual event of sheep ranchers herding thousands of sheep into the mountains for summer grazing, then back down for shearing and/or transportation to animal processing plants. Filmed in epic long shots that last for over a minute, sometimes more than three minutes, and with sync sound, *Sweetgrass* immerses the audience in what Castaing-Taylor has called a world of "sheeple": "*Sweetgrass* is interested in both the sheep and the people, or more precisely their intertwined naturecultures within the context of their larger ecological fold".[12]

The focus of the sound in *Sweetgrass* is non-linguistic utterance: the bleating of the sheep most obviously, but also the many non-verbal utterances of the ranchers who shear, breed, and herd the sheep into the mountains. Though the film is sync sound, moments of what would normally be called dialog are relatively rare and function as a kind of punctuation within the ongoing depiction of the naturecultures of the sheep and the cowboys and their dogs. Further, sound and image work together synergically to expand our sense of the ecological surround within which the business of raising sheep for wool and meat takes place. In multiple instances we hear in intimate close-up precisely what we are seeing in extreme long-shot.

Leviathan, the first collaboration between Castaing-Taylor and Paravel, also uses time-extended shots, but in a radically different way from *Sweetgrass* and from most of the films discussed in this chapter. In *Sweetgrass*, Castaing-Taylor, who shot most of the film, is careful to be sure the viewers' focus is on the sheep and sheep-herding, not on himself and his filming. *Sweetgrass* is sonically, and from time to time visually, immersive, but Castaing-Taylor himself is more observer than character. The natureculture of the fishing boats Castaing-Taylor and Paravel shipped out on from New Bedford, Massachusetts, made any attempt toward visual detachment impossible: *Leviathan* is, from its opening moment, fully immersive both sonically (seen theatrically, the film is surround sound) and visually, for both the audience and the filmmakers themselves, whose GoPro cameras record the movements of ocean, boat, fishermen, and filmmakers as they struggle to get their work done in an entirely unstable world. Indeed, the continual motion of ocean and fishing boat and the filmmakers' cameras make determining where shots begin and end pointless. The experience of *Leviathan* is less about patience and a more thoughtful meditation on nature and/or the interplay of the natural world and the human world, than about stamina within an immersive and unsettling experience for the filmmakers and their audience. In the film's credits, Castaing-Taylor and Paravel list each individual human and each specific species of fish, shellfish, bird, seen and heard, together, reconfirming the implication of *Sweetgrass* that nature and human culture are not separate arenas—and implicitly demonstrating their respect for both the power of the ocean and the men who drag a living from it.

Sniadecki's *Songhua* is named for the Songhua River, which flows from the Changbai Mountains through Manchuria, joining the Amur River to form a natural boundary between China's northernmost province, Heilongjiang, and Russia. Sniadecki arrived in the city of Harbin, on the banks of the Songhua, not long after news that serious pollution of the river had occurred. The three extended shots that open the film (93 seconds, 132 seconds, 59 seconds) create a contemplative pace that allows Sniadecki to develop a panoramic sense of goings-on in, on, and around the river. Each shot reveals a particular activity—men throwing fishing nets into the river, people walking on a dock near where a man is grilling meat, a woman with a loudspeaker promoting tour boats—within a larger context so that, during the film's 29 minutes, a network of visual and auditory interconnections is created between the different views, enhancing our sense of this small but complex environment and its recent environmental history.

For Sniadecki, the Harbin waterfront is an engaging and fascinating space, but also representative of the ways in which industrialization and commercialization throughout the world inevitably affect the natural environment—as well as of the way in which environmental damage is usually processed by those who hear about it. Sniadecki's apparently effective Chinese (others in the film are surprised by his agility with the language) allows him to talk with individuals who reveal mostly complacent and hopeful responses to the fact that nitrobenzene from a processing plant had recently polluted the river. *Songhua* is an emblem of our own ways of evading the implications and demands of a changing global environment.[13]

Manakamana was filmed in Nepal on a modern cable car installation that had been imported from Austria, offering easy access to the mountaintop temple of the goddess Bhagwati—long a pilgrimage site that, until the cable car installation, took several days of climbing on foot to reach. Shot in 16 mm, implicitly as an act of nostalgia by the filmmakers, *Manakamana* records eleven rides on the cable car, six up and five down. For each ride, the camera and filmmaker are located on one side of the cable car and film single-shot portraits of the individuals (several of them familiar to Spray from her earlier ethnographic research and earlier films) who sit across from them. Each shot lasts as long as the cable car ride, around ten minutes. As the trips accumulate, we see both the travelers and, through the windows of the cable car, views of the cable car apparatus (sometimes other cable cars pass by), as well as bits of the surrounding terrain. Spray and Velez film from both sides of the car so that the view from the windows reveals the landscape and other passing cable cars in four different ways. Though there is minimal interplay between Spray and Velez and the travelers (once, a woman smiles at the camera), the travelers often talk about what they're seeing out of the windows and how the installation of the cable car has changed the experience of visiting the temple.

Manakamana is a meditation on ways in which modern life continues to transform tradition. Whereas the travelers they film are focused on honoring the goddess Bhagwati or at least visiting a famous spiritual site, Spray and Velez do not visit the temple, though we hear the temple bells and the bustle of the pilgrims during an extended moment of darkness midway through the film, dividing the trips up from the trips down. The overall structure of the film echoes Benning's use of the 10-minute shot in *13 Lakes* and *TEN SKIES*.[14] In *Manakamana* the filmmakers and their subjects are literally enclosed within technology, and their ecological surround is evident to viewers of the film, only in a limited way, sometimes through the cable car windows and through descriptions by some of the visitors of how the surrounding terrain has changed over the years (Fig. 3.3).[15]

Coda

Traditionally, we have understood ecology as the study of relationships between living organisms, including humans, and their physical environment. In recent decades, we've come to recognize that our ecological surround is more than the living world, that we and all other living things function within a continually expanding media ecology. Among the most unusual recent films that use extended single shots as a means of creating what is, at least implicitly, an ecological meditation—in this case, during which we never have a glimpse of the world outside our containment within a technologically created space—came to my attention when I received an email in August 2020, from Jennifer Anderson and Vernon Lott about their film *#monalisa* (2020). They explained,

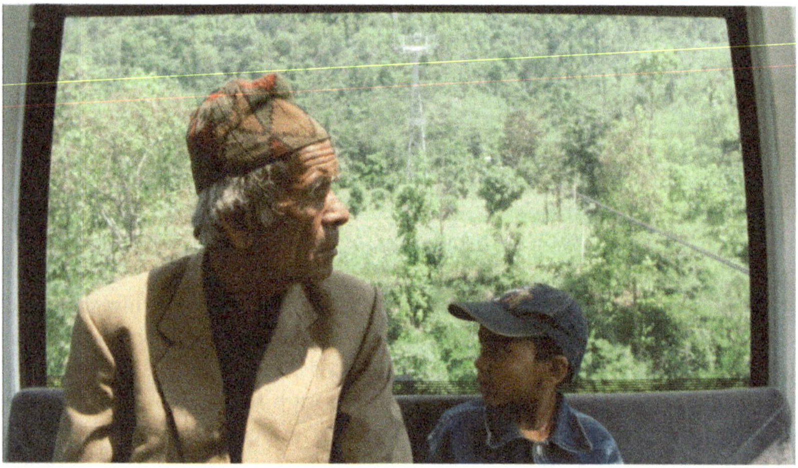

Fig. 3.3 Stephanie Spray and Pacho Velez, *Manakamana* (2014)

> [In] a world where our attention spans are waning (a 2013 study conducted by Microsoft found that the average person's attention span was eight seconds—one second less than that of a goldfish), *#monalisa* invites viewers to slow down and observe.... The film is a contemplation of human behavior and a meditation on looking.... At eleven hours...it's not necessarily meant to be watched from beginning to end. Viewers can click around on the link and watch it in any order.[16]

When I accessed *#monalisa* on-line (it's available at Anderson/Lott's Morris Hill Pictures website), I discovered that its subject is the room in the Louvre where Leonardo da Vinci's *Mona Lisa* (1503) is exhibited to, and guarded from, throngs of people from around the world, many of them determined to record a selfie of themselves with the painting. My curiosity led me to watch the first hour of the film, which led to my watching the entire film as a daily one-hour meditation (the score by composers Peter Broderick and David Allred complements the film's meditative visual approach).

Not only is *#monalisa* made up of extended shots, sometimes longer than 30 minutes (though mostly shorter: 5, 6, 10 minutes), but all the shots are in slow motion. Nandan Rao agreed to film in the *Mona Lisa* room and bought a GoPro camera and several accessories. The idea of shooting in slow motion was a result of Rao's discovery that the GoPro could film at 120 frames per second; he made a few test shots, and Lott and Anderson agreed that slow motion was best for the situation. The experience of *#monalisa* evokes the earliest experiments in motion study photography by Eadweard Muybridge and Etienne-Jules Marey.

At first, witnessing the crowds of adults squeezed together to see a painting they cannot stand close to (a complex set of fences, plus museum guards, keep the crowd at bay) and the fact that they seem less interested in actually seeing the *Mona Lisa* (more easily and completely viewed merely by googling the painting) than in being seen with it provides absurdist humor. But if one gets past first reactions and pays attention to the slow-motion activities captured by Rao, the kinesics of the individuals in the crowd, including Rao, become fascinating to consider. Muybridge analyzed individuals-in-motion; Anderson/Lott/Rao allow us to truly *see* a modern crowd-in-motion during a particular historical moment and in a place where what is most fully natural is the human bodies in the space and their interrelationships with one another as they move.

#monalisa can function as a metaphor for the project this chapter is part of. Those of us who write about the ways in which cinema, including what we have usually called "avant-garde cinema", has taken account of our ecological surround do this work by screening films in our home or institutional screening rooms in order to study them and by sitting at our computers to write. The films we study and discuss are our Mona Lisas and our focus is our own "selfies", our responses to the films. That the particular kinds of slow cinema I've discussed here have evolved in the United States, the world's most consumption-oriented and wasteful nation, is, of course, an obvious eco-cinematic irony; and even the estimable films I've discussed cannot help but participate in the pattern of resource exploitation, over-consumption, and waste that are transforming the global ecology. Emulsion-based cinema was/is a dirty and often wasteful medium and we are seeing how quickly the systems and hardware that enable digital production, distribution, and presentation are rendered out-of-date, becoming a new category of waste.

At most, the single-shot eco-cinema I've described is a cinematic canary in the coalmine of a media-saturated world, little more than a series of quixotic gestures—or prayers, perhaps. The idea that these films might make a substantive difference, given our current ecological challenges is, if the reader can forgive the pun, the ultimate long shot. Nevertheless, we must continue to hope that, decades hence, this eco-cinematic long-shot cinema, and our efforts on its behalf, will have come to be recognized as a premonition of what will have become a more aware and responsible human engagement with a transforming global ecology.

Notes

1. A revised version of the essay was published in Sean Cubitt, Salma Monani, Stephen Rust (eds), *Ecocinema Theory and Practice*, an AFI Reader, New York/London, Routledge, 2013, pp. 17–41.
2. Geyrhalter, *in* Scott MacDonald, *The Sublimity of Document; Cinema as Diorama*, Oxford, Oxford University Press, 2019, p. 152.
3. It is made clear in this series, and confirmed in *David Attenborough: A Life on Our Planet* (2020), that climate change is a result of human activities. At the UN Climate Change Conference in 2018, where he predicted disaster for the planet if humanity doesn't make major adjustments, Attenborough showed some of this walrus footage to the audience.
4. There is an irony here, of course: the Lumière Brothers were businessmen interested, at first, in producing as many screenings as they could, and, in the long run, in selling the Cinématographes they manufactured.
5. Larry Gottheim, in Scott MacDonald, *A Critical Cinema*, Berkeley, University of California Press, 1988, pp. 84–85.
6. In 2020, Re:Voir, in conjunction with the New York Filmmakers' Cooperative, released a DVD of Gottheim's early films up through *Barn Rushes*: "Larry Gottheim *Fog Line*". The version of *Fog Line* on the Re:Voir DVD is far superior, particularly in its representation of the lovely green colour of the original 16 mm prints of the film, to the version included on "American Film Treasures IV: Avant Garde Film, 1947-1986", released by the National Film Preservation Foundation in 2009.
7. Hutton, in Scott MacDonald, *A Critical Cinema 3*, Berkeley, University of California Press, 1998, p. 243.
8. For many years, Benning has taught a course called "Looking and Listening" at California Institute of the Arts.

9. Language spoken by the Shoshone Indian tribe of North America, who occupied a territory between southeastern California, central and eastern Nevada, northwestern Utah, southern Idaho and western Wyoming.
10. J. Benning, *in* Scott MacDonald, *Adventures of Perception*, Berkeley, University of California Press, 2009, p. 259.
11. In *Goshogaoka*, *NŌ*, and *Double Tide*, the activities we see seem at first to be fly-on-the wall documents, but in all three films it soon becomes clear that what we are seeing has been carefully choreographed by Lockhart. This is quickly obvious in *Goshogaoka*, much subtler in *NŌ*, where we gradually understand that the farmers are mulching *only* that section of the field visible in Lockhart's image. Similarly, in *Double Tide*, Casad carefully, and obviously, never leaves the frame. The continuous image in *Teatro Amazonas* is accompanied by a chorus, apparently on stage and watched by the audience, singing a single note, though gradually the number of singers becomes smaller until we only hear the sounds of the audience itself. Of these films, *NŌ* is most relevant here. See Lockhart's comments in *A Critical Cinema 5*, pp. 328–331.
12. Castaing-Taylor, in Scott MacDonald, *Avant-Doc: Intersections of Documentary and Avant-Garde Cinema*, New York, Oxford University Press, 2015, p. 383.
13. Sniadecki would continue to explore the long single take, most obviously in his collaboration with Libbie Dina Cohn, *People's Park* (2012), a 78-minute single-shot document of a public park in Chengdu, China. Less obviously meditative than the other films discussed here, *People's Park* provides a view of a diverse Chinese crowd engaged in a wide variety of holiday activities, stripped of conventional cliché and political commentary.
14. Both Spray and Velez had studied Benning's work: Spray at the SEL, where Benning's films were regularly shown, and Velez at CalArts, where he studied with Benning. Spray: "[Benning's] films made a strong impression. *Thirteen Lakes* was the first that I saw, and is a favorite", in MacDonald, *The Sublimity of Document*, p. 468. Spray once admitted to me that at least one reason for using six trips up to the temple and just five down was to avoid seeming too indebted to Benning and the often mathematical structuring of his films.

15. While the windows of the cable car function as frames within the film frame, the camera is always pointed at the cable car riders, never out of the window. I see this as an echo of how in *11 × 14* Benning suggests narrative, but refuses it. Spray and Velez refuse viewers a standard pleasure of a cable car ride, the expansive views out of the windows.
16. A more complete version of this email is included in Scott MacDonald, "Visiting the Louvre with *#monalisa*: An Interview with Vernon Lott, Jennifer Anderson, Nandan Rao, Peter Broderick, and David Allred", *Cineaste,* Vol. 46, no. 2, Spring 2021, p. 22.

CHAPTER 4

The Disvertebrated Cinema

Jean-Michel Durafour

Arthropod Didascalia

The master knows the art of proportions and the relationships of forms. He learned it in the family workshop run by his older brother, Goessen, in s-Hertogenbosch, Den Bosch, the hometown to which he owes his nickname.[1] And yet, in the *speculum nuptiarum* of *The Garden of Earthly Delights* triptych (1494–1505), which is perhaps no more than a database of emblems and motifs for moral edification, and for which he had to travel to the court of Nassau despite being a notorious homebody, he introduced a tiny detail which most viewers do not notice, like a secret in paint hidden in the thickness of the canvas, barely visible, recognizable as a butterfly of no significance, whose representation undermines both the thesis of a resurgence of ancient visual formulae and that of the emphasis on realism in the recently invented Italian new painting. For, at the edge of the pond, the wings of this lepidoptera, very common at least before the age of carbamates and rotenone, an *Aglais urticae* or small tortoiseshell, foraging from a blue thistle in the central panel, and

J.-M. Durafour (✉)
University Aix-Marseille, Marseille, France
e-mail: jean-michel.durafour@univ-amu.fr

E. Della Noce and L. Murari (eds.), *Expanded Nature*, Experimental Film and Artists' Moving Image,
https://doi.org/10.1007/978-3-031-70729-2_4

whose figurative event is out of all proportion to the pair of wings of a *Maniola jurtina* mounted on a bird's body a little further away, these wings, folded vertically one on top of the other, are here represented with the designs and colours that are normally found on their ***dorsal or top side***, the very ones that the closed wing here hides, as if the undersides of the painting were here exceptionally visible, the whole painting being perhaps no more than a visual fold, a "work devoted to converting the organic detail (something that is ***in the body***) into a true morphogenetic place (something ***in which the body is formed***)".[2]

The painting—which I have often revisited—is a ***dispositif*** that must indeed be opened to see. The triptych is already in the butterfly, the copy, in the model; one could talk for pages and pages about the theoretical exaltation of such an image.

*

Let's now turn our gaze to this young university graduate who has just moved from Nishijin to Shimogamo to continue his research on riparian insects. We are in Kyoto in the early 1940s and Imanishi, who is not yet the famous primatologist he will later become, is particularly interested in the classification of the mayfly taxon on the banks of the Kamo River. Without going into too many meandering details about the exact nature of species (real or just nominal entities?) or the place of Imanishi's studies in the legacy of biological thought (Wagner, Jordan, Grinell), one important point can be retained from his work: depending on the speed of the current, four kinds of mayfly nymphs of the suborder *Schistonota*—namely *Epeorus uenoi*, *Epeorus curvatulus*, *Epeorus latifolium* and *Ecdyonurus yoshidae*—divide their habitat between themselves from the river's axis to the bank. Contrary to Darwin, for whom it plays a very marginal role, the biotic environment is so crucial here that it leads Imanishi to reject the traditional term "species" in favour of a neologism that places greater emphasis on the medial integration of individual beings: "ecospecies [*sumiwake*]", in the translation given by Augustin Berque, literally meaning "to inhabit [*sumu*]"/"to separate [*wakeru*]". This is "the concrete substance of what a living species is", or what Imanishi calls "the living form of an animal species"[3]: animal and habitat are inextricably linked and evolve together in such a way that "the real living thing is an integration of the living thing and the environment on which it depends and ***that is the organized system of the living thing itself***".[4] The same applies to images, if we accept that they can be apprehended according to the model of a living being and under certain conditions.[5]

This is not to say that images cannot be dissociated from their more or less immediate environments (which is self-evident), whether artistic, political, economic or religious; nor that we are interested in the contexts in which they appear or the plurality of space-times in which they are received, etc.; but rather to affirm that an image is only ever the expression of an ecological co-participation between the image and its environment (as early as 1977, Sontag had already spoken in this sense of an "ecology of images", but this idea was being applied long before the word existed: this was the Warburgian project against formalist aesthetics from the outset). To put it another way: images are not things but relations.[6]

*

The physiologist's attention is drawn to just one aspect of human movement: its figurative animal matrix. With Étienne-Jules Marey and Eadweard Muybridge, a new visual and anthropological pact was sealed, which was to culminate in the cinematographic image and animal bodies: we are now only interested in man in terms of animal locomotion. If—as Béla Balázs wrote in the 1920s—indexical cinema put an end to the centuries-long cultural supremacy of the *legible* that began with Gutenberg's invention of printing, and finally made man *visible* (painting only ever showed us his style), it was through a humanism that kept below the human. The animal makes the human visible. Descartes believed in the "animal-as-machine" (although he never uses this expression *stricto* sensu); cinema was to be the *machine-as-animal*.[7] In 1950, in "The Fluid World of the Screen", Jean Epstein compared the camera to the eye of the snail, mounted on a retractable stem, or to the multifaceted eye of an insect. Since then, its animal dimension has been constantly emphasized: two *inhumanities*—of the animal, of the machine—unite to launch the new perception addressed to human beings by filming, editing and projection devices.

The historic invention of the Lumière Cinematograph is first and foremost a redeployment of Ovidian metamorphoses—where human and animal anatomies are not distinct in the world frequented by the "gods" (the name given by Jean Louis Schefer to the gigantic, mutant beings bustling about on the screen in his 1980 essay *The Ordinary Man of Cinema*)—through the prism of recent Darwinian evolutionary theory. From the outset, this gap-bridging takes the form of an unexpected pact between the sciences and the arts. Cinema confronts us with visual evidence of the plastic *continuum* of forms and the instability of species ever since the "primal protoplasm, not yet possessing a 'stable' form, but capable of assuming any form and which, skipping along the rungs of the evolutionary ladder, attaches itself to any and all forms of animal existence".[8] Darwinian gradualism did not invent biological thought-by-metamorphosis (Lamarckian transformism came first), but henceforth all is generalized metamorphosis, to the point where similarity, which defined how older taxonomies were classified, is no longer qualified to organize the living. It is no small irony of fate that Darwin's ideas rapidly gained acceptance throughout Europe, with the notable exception of France. They took on a new twist there.

Monkeys invented cinema. The logic of the *animated* image is, first and foremost, the figurative certainty that we are not separate from animals and that there is nothing but an incessant and transitory coming and going between all living beings. This last point is important. Epstein—him again—never stopped repeating this: the distinction between nature's realms is ultimately just a matter of image speed (in fast forward plants move like animals; in very slow motion, humans return to their "fundamental viscosity"[9]) and it is only in terms of our eyes' capacities that certain portions of reality are dissociated. Cinema is also a democratic art in that it cancels out the conventional hierarchy of nature's kingdoms; it is the theoretical instrument of a flat biology. The images of cinema are alive first and foremost because they treat all things as living beings, on an equal biological footing.

The camera cares little for human vision. Historically, cinema—and others have joined it— has been the most advanced machine for leading us to animal sensation. For there is nothing natural about the animal. Something like a natural state simply does not exist: how far back exactly would we have to go to find it? The animal is on the other side of human manipulation. Every animal is a machination.[10] The animal is ahead of us. The fact that cinema has progressively imposed itself at the very moment when the animal has increasingly disappeared from our daily lives shows the extent to which cinematographic images and expressive animalities can be substituted for one another.[11]

Cicindela-Images: Speed, Blindness, Figuration

Here, then, is a play—or a ploy—with three characters: the insect, cinema and the ecology of images. In this scenario, it's easy to see that it is not all the transitions from human to animal for which cinema has been a major source that are important, but rather the entomological metamorphoses of the image itself. The insect is here another name for visual experimentation in the cinematic image.

On the River Tisza, a tributary of the Danube, a young Serbian girl is confronted with the spectacle of the efflorescence of mayflies, from their first aquatic appearance to their death on the surface of the current. If the title of this forty-minute medium-length film—*Vita brevis* (2014) by Thierry Knauff—partially quotes a medical formula attributed to Hippocrates, which is specifically human (all the philosophies of detachment and hedonism also converge in it), the black-and-white images divert it by bringing together several forms of biological life between which they establish certain relationships and whose centre of gravity they shift towards the animal as the eye gets closer to the smaller beings: between fragile human life and the short programmed existence of insects in general (a moment that, with C. S. Peirce, we could call symbolic), between the youthfulness of the little girl and the hatching of adult insects in a particular narrative moment (a moment of analogical resemblance), between the two *metamorphotic* stages of the adult mayfly when the gaze focuses on the animal world alone (a documentary moment of physical contiguity). Indeed, to start with this last point, the *Ephemerica danica* has the property of having two adult stages instead of just one. Once past the egg-hatching stage, which is of no interest to the filmmaker, the mayfly first experiences an aquatic larval destiny (the "nymph"), before spending some time on dry land (the "subimago": the wings unfurl and dry out), then developing into a final flying and sexually mature form (the "imago"). Unlike butterflies, mayflies do not have a chrysalis (the subimagos are mobile), but rather a continuous series of moults. Moreover, while mayflies are not the only insects with subimagos, they are the only ones that have another special feature: they first appeared in the Carboniferous period and are the oldest winged insects still around today—so the title of the film sounds like an antiphrase. In other words, while the order of Ephemeroptera has not undergone any morphological modifications over several million years, its individuals undergo the greatest number of anatomical transformations in a record time of just a few hours. This slowing down of long time and this acceleration of short time is quite remarkable. By threatening the mayflies, which are highly sensitive to chemical and light pollution (at one point we see a crop-duster plane whose invasive flight is mirrored by the graceful flight of a heron), human beings are undermining this balance by conditioning their regression: they are short-circuiting the species' slow time. Without using the slightest visual effect—such as slow motion, time lapse or freeze frame—the film returns the human being to an animal logic by *the very fact of the cinematic image*. Human beings' brief lives are nothing more than the on-screen extension of neuropteran avatars. The shots in which the young girl, in the prime of her life, is caught in the middle of the flight of mayflies as they blossom only to die, are striking in this respect. Without any showy or demonstrative aids, Knauff achieves a masterfully figurative demonstration: faced with the common destiny of death, the human being, at the other chronological and evolutionary end of all this animal old age in nature, is an insect like any other, and it is shots like any other that are responsible for depicting it. Immature, with an infantile body, the girl is a subimago who does not *yet* have operational wings. That will come. Such is the animal connection of the cinematographed bodies to one another.

And the film can do this because, in nature, the insect is and always has been a cinematic being.

That is to say, the insect is a *vector of image modulations that chases the human being from our gaze and makes us see as animals*. Bosch is not the only one to have staged this very early on. Since the anatomical plates in the first entomology treatises of the Renaissance and the Classical Age—Thomas Muffet's *Theatrum insectorum* (1589), Ulisse Aldrovandi's *De animalibus insectis libri septem, cum singulorum iconibus ad vivum expressis* (1630), Joris Hoefnagel's *Diversae insectarum volatilium icones ad vivum accuratissime depictae per celeberrimum pictorem* (1630)—the insect is not so much an objective animal as an experience of vision inviting the human being to abandon its anthropic eye for regimes not only of irreverence but also of "irreference"[12] by which the reader-viewer was asked to situate themselves somewhere between their private consciousness, sheltered in the dark room of their cranium,[13] and the place of the magnified insect, in a double test of object magnification and subject "shrinking". Driven out of myself, I can finally see what my ordinary perception neglects because of its smallness or speed, and which I can only observe properly if I expel the living from it.

This is also the situation in front of the cinema screen, where I am invited to occupy the place of the insect (I look at them as they would see me), but of a reinvigorated insect (the images move, I react to them). It is perhaps experimental cinema, or "experimental-leaning" moments in narrative and representational films (for example in Vera Chytilová, Jerry

Lewis or Seijun Suzuki), that in the history of moving images best isolate this departure from human perceptions (always acquired and cultural—nothing exists as a natural perception, not even and especially not animal perceptions[14]), by highlighting the pre-cinematographic (in the sense of the historical *dispositif*) figurativity of insects as a *mise en abyme* in the image of filmic procedures. The insect is not so much filmed as *filming*, in that it renders cinema's figurative operations visual—if not quite visible—within the image.

As it is beyond the scope of this text to explore one by one the different theoretical modalities of cinematic perception's entomological register, I shall confine myself to describing one of the most important in some detail: the question of movement detached from figurative mobile bodies, which we know to be the great visual project of cinema since Marey's chronophotographic decomposition abandoned "the consideration of the stationary points occupied by the moving body ... in favour of the evanescent transitions that connect them and whose visibility has to be expressed on the basis of a set of visual formulae that do not presuppose figurability, but on the contrary condition it".[15] We can distinguish three major speculative dimensions of the filmic image's movement: from projection on the screen (extrapellicular), the movement of the perforated strip (pellicular), the movement inscribed in each frame (intrapellicular—cinema is not animated photography, every frame contains "the beginning of movement, *blurred signs*").[16]

*

Paolo Gioli's *Farfallio* (1993)—the title of which is a portmanteau of butterfly (*farfalla*) and flicker (*sfarfallio*)—associates the fluttering of butterfly wings and the opening of the whirring projector that projects and animates film images onto a screen (Fig. 4.1).

By spreading their wings, butterflies reveal the top (most aesthetic) side, then withdraw it from all possible view by folding them over each other: butterflies already cinematograph the visible. The images of the spectacle appear in the opening and disappear in the closing. The insects—whose furry bodies are also likened by the filmmaker to women's genitalia in pornographic sequences that overlap with images of the lepidoptera—are taken from photographs borrowed from books and albums, filmed frame by frame then reanimated by editing and projection. Érik Bullot recently detailed in a very beautiful text some of the technical processes involved in the junction between entomological morphology and cinematic invention:

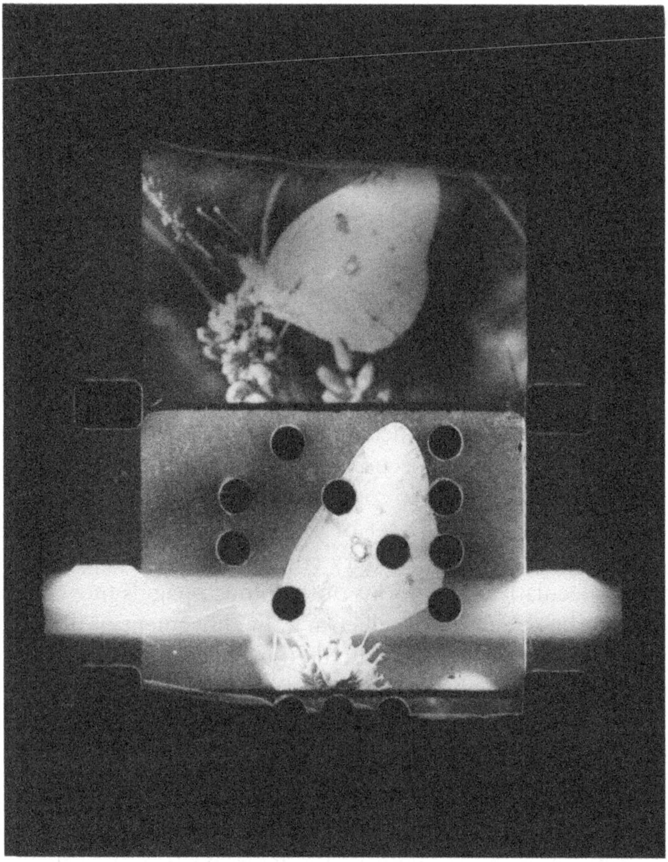

Fig. 4.1 Paolo Gioli, *Farfallio* (1993) © All rights reserved by the artist/ Courtesy of Light Cone

> … The difference between white and black butterflies, which creates a graphic flickering effect, like a signal, recalling the opposition of positive and negative images … ; the symmetrical separation of the image produced by the two wings of the butterfly, evoking the principle of the *split screen* … ; the eyespot likened to a look of terror or fascination by the superimposition of human looks taken from classical painting or filmed portraits.[17]

In this eight-minute film, Gioli's intention was in his own words to "be ironic"[18] about the flicker of the clumsy technical beginnings of proto-cinema, later deliberately incorporated into experimental cinema.

*

The identification between insect and cinema is taken a step further in *Mothlight* (1963) by Stan Brakhage, a four-minute short film made without a camera or traditional filming but by direct intervention on the film strip. Brakhage mixed butterfly wings, flower petals and blades of grass, placed them between two layers of transparent polyester film, forming a film strip, and then ran the whole thing through an optical printer. His gesture is an extension of the various artistic experiments with film as material (Man Ray, Len Lye, Isidore Isou), but also of Jean Dubuffet's entomological work on pictorial matter between 1953 and 1955 (*Belle au regard masqué*, *Le Strabique*), where the discs, veins and scales of wings allusively trace the details, shapes and complexions of faces, and—unlike Picasso's *Composition au papillon* (1932) where the lepidoptera, integrated into the collage, signified nothing other than a butterfly—serve to question the capacity of matter to create form: in these canvases, as Michel Thévoz notes, "[the butterfly] positions itself as a signifier, it redirects the questioning of the gaze and its demand for reality, it replaces the paste to prevent the image from clogging the imagination".[19] It "reminds the eye of its perceptive responsibility".[20]

Brakhage's "herbarium through transparency [*herbier par transparence*]", to use Érik Bullot's expression,[21] this "stained-glass film [*film-vitrail*]",[22] not only elevates the insect to the rank of raw material for the film, but above all makes cinematic movement the latest avatar of the butterfly, the scrolling of the projection adding a final wing-flapping to the image-imago. Rather than an animation film, we are dealing here with a *reanimation film*. We know that Brakhage developed a whole experimental theory of vision that he called "*closed-eye vision*"[23] (after the autodeictic phenomenon of phosphenes). In optical phenomena like myodesopsias, which form on the inner face of closed eyelids, the image no longer comes from external forms, but—bordering on abstraction (spots, filaments)—is produced by the eye's own light and comes from inside the body. Let's not forget that cinema was born at a time in the nineteenth century when the old model of the darkroom had given way to the new physiological science and to the return of vision into the body, in particular through the observation of the same visual events ("The hole [in the window of the dark room] being then closed, let him look towards the darkest part of the room; a circular image will now be seen to float before him. The middle of this circle will appear bright, colourless, or somewhat yellow ...").[24] With *Mothlight*, eye "flies" (called *mouches volantes* in French and *muscae volitantes* in Latin, meaning "flying flies"), insects and cinematic energies come together in an image fomented with living material and all the world's glimmers.

*

In *De-Extinction* (2014), a colour film lasting just under fifteen minutes made with micro cameras and macro lenses, Pierre Huyghe records the long eternity of prehistoric insects trapped in amber. When thinking of this type of entomological body as a matrix of figurative invention for cinema, the fact that this was perhaps at the heart of Steven Spielberg's *Jurrasic Park* (1993) comes to mind. In the same way that researchers isolate the incomplete genomes of certain dinosaurs in the bodies of mosquitoes preserved in petrified resin, it is also through the intermediary of another mosquito, this time a cartoon character, that the spectator and incidentally the park visitor is presented with the process by which certain antediluvian species, *Tyrannosaurus rex*, *Velociraptor*, *Triceratops*, have been brought back to life. The sensational special effects park is none other than the eponymous film—as evidenced by the famous scene where a computer screen displaying the dinosaurs' DNA sequences, and referring in the image to the digital coding of the film's CGI artists, is reflected on the body of a velociraptor. It's a clever demonstration, because while the film's *narrative* insists, as a popular entertainment endeavour, on the photorealism of digital anatomies (of which there are ultimately very few), the same cannot be said on a *theoretical* level, insofar as animation and CGI have in common the fact that they are not based on photographic recording, as opposed to indexical cinema.

Digital cinema, at the other end of the chronology, joins up with the drawings of "prehistoric" cinema (Plateau, Muybridge), by folding the time of cinema itself. Analog film, for its part, escapes the grip of the mosquito, which it keeps captive in its amber (the knob of the billionaire's cane, a photographic insert in the animated sequence)—a biting irony when one thinks of all the miniature insects that no camera has ever been able to prevent from imprinting on their shots and whose power of auto-fictionalization through their irruption into the image field according to the disparate whims of natural contingency is pointed out by Julien Gracq, in his *Reading Writing* [*En lisant en écrivant*] (1980).[25]

In both Huyghe and Spielberg's works, the fossilized insect is an opportunity to make a visual statement about the nature of analog cinema. In *De-Extinction*, the tiny details of an amber stone are captured by the camera in slow close-ups whose deliberately blurred focus disturbs the scale of proportions. A fantastical landscape gradually takes shape, until the amber is revealed when the camera pulls back. Prisoners of their gangue (the ancient Greeks called amber *êlektron* because its rubbing produces static electricity), small beings several million years old—we are confronted once again, but in a different manner to *Vita brevis*, with this state of living things stuck in a very long geological time—act as ghostly witnesses in the image and offer a figurative reflection of the film medium's status as "change mummified".[26] As Jean Louis Schefer has written, cinema has accelerated the ageing of time[27] and Huyghe makes cinema the stigmata of what it will never have been able to record and the harbinger of a new visible, perhaps post-apocalyptic world to come, where flesh and stone will no longer be entirely distinct (the film is set against the backdrop of the new mass extinction of animal species for which man is largely responsible, and therefore also of his own demise). Expiration is not necessarily an end but, as in breathing, can be followed by a new breath, a new cycle of life. The term "de-extinction" was initially introduced to designate behaviours aimed at reversing the extinctions of living beings by restoring new versions of extinct species to their former habitats.[28] *Jurassic Park* is decidedly not so far off.

It seems in a sense then that it was inevitable that the powers of film would set mummified insects back in motion. This is what happens in Momoko Seto's video *Planet Σ* (2014) where, for about ten minutes, we can observe insects, bees, cockroaches, orthoptera, being released from their ice coffins in fast forward. There is a familiar Arctic imaginary at play here: having served for over a century as a gateway to a "lost world" (Jules Verne) or containing specimens of extraterrestrial life (Lovecraft). In the intergalactic reaches of the universe, Σ is a planet of giant insects filmed in microphotography, which in a way follows on from Pierre Huyghe's film: life resumes its course on a star close to the early formative times of our own (Seto's planets are made up of the four elements: gas, volcanic eruptions, etc.), where humans have disappeared and where their absence allows the return of vanished species. This rebirth is once again accompanied by a disruption of time: in the same shot, Hymenoptera fly in slow motion over a mushroom growing at tenfold speed. From one shot to the next, the time of the image is frozen in the act of creating a world.

Somewhere between naturalism and science fiction, Seto uses insects to propose inhuman modules of perception, which the director is well aware are still threatened by the obstinacy of human practices (a final shot shows dead bees that have fallen to the ground).

*

To say that cinema takes on the posture of an entomological perception in these few examples is certainly not to speak of a perception that would mimic this or that variety of insect. Rather a perception *in place* of the insect. Through the animal resolutions of an equipped *dispositif*, human perception takes the place of that of the mosquito or the butterfly without seeking to imitate them—which is why it is entomological, following an insect logic, but is not insectoid, morphologically similar to that of the insect—which would be quite impossible and has only ever given cinema some quite fanciful opportunities (like the famous kaleidoscopic vision of compound eyes in Kurt Neumann's *The Fly* [1958]: it is futile to want to see *like*).[29] To see in place of the insect is to affirm that we see from its place but also in its place. And experimental cinema allows us to do this better than anywhere else, since it shares with science fiction (or more generally fantasy—the other major theoretical site of insects in films) the use of our vision as an object of our decentred human gaze.

Notes

1. The present article is composed mainly of original, material, however, some sections are slightly modified paragraphs extracted from the following texts already published: "Le cinéma à l'usage des vivants: Réflexion après une lecture de Bruno Latour", *Débordements*, 2 June 2021, https://www.debordements.fr/Le-cinema-a-l-usage-des-vivants; "Cinéma et papillons encadrés. Petit manuel d'entomologie filmique", *Débordements*, 7 January 2020, https://www.debordements.fr/Cinema-et-papillons-encadres.
2. Georges Didi-Huberman, *Phasmes. Essais sur l'apparition*, Paris, Éditions de Minuit, 1998, p. 195. Unless otherwise noted, all translations from the original French are by the translator.
3. Kinji Imanishi, *La Liberté dans l'évolution: Le vivant comme sujet* [1980] (followed by "La mésologie d'Imanishi" by Augustin Berque), trans. Augustin Berque, Marseille, Wildproject, 2015, p. 82 for all the citations in this sentence.

4. Kinji Imanishi, *A Japanese View of Nature. The World of Living Things* [1941], trans. P. Asquith, H. Kawakatsu, S. Yagi, H. Takasaki, London, RoutledgeCurzon, 2002, p. 60. [NB: This sentence in the French version of Imanishi's text is more succinct and radical: "the living being and its environment are one, *and that itself represents the real concrete living being*". Kinji Imanishi, *Le Monde des êtres vivants: Une théorie écologique de l'évolution* [1941], trans. Anne-Yvonne Gouzard, Marseille, Wildproject, 2011, p. 82.].
5. Everything depends on the meaning we give to the word "life", which has long since ceased to be the ambition of biology itself, which generally only retains a pragmatic usage and a chemical clarification, i.e., specifically based on the non-separation between the inert and the animated. As Imanishi also writes: "I do not find it a problem to admit the life of nonliving things", *ibid.*, p. 19. Questioning the life of images is one of the main objects of my work *Cinéma et cristaux. Traité d'éconologie*, Sesto San Giovanni, Mimésis, 2018, to which I refer the reader for any desired clarification. The present text takes its conclusions as given.
6. One of the interesting consequences of this situation is what I have called iconicization. See Jean-Michel Durafour, "Métastases. Croissance cellulaire des images et film rival", *Débordements*, 4 April 2019, https://www.debordements.fr/Metastases.
7. It is worth remembering that in 1873 Marey in fact published a work entitled *La Machine animale* [literally "The Animal Machine", but translated in English as *Animal Mechanism*], which was a transition between his early medical writings and his physiological work on motricity.
8. Sergei M. Eisenstein, *Eisenstein on Disney,* ed. Jay Leyda, trans. Alan Upchurch, Kolkata, Seagull Books, 1986, p. 21.
9. Jean Epstein, *The Intelligence of a Machine* [1946], trans. Christophe Wall-Romana, Minneapolis, Univocal Publishing, 2014, p. 29.
10. We know nothing about animals outside of a consideration by the human being. *All* animals are zootechnical inventions of the human being. On the case of the wolf, for example, see Baptiste Morizot, *Les Diplomates: Cohabiter avec les loups sur une autre carte du vivant*, Marseille, Wildproject, 2016.

11. See Étienne Souriau, "L'univers filmique et l'art animalier", *Revue internationale d'esthétique*, no. 25, January–March 1956.
12. I borrow these two terms from Dominique Noguez, *Éloge du cinéma expérimental*, Paris, MNAM/Centre Pompidou, 1979, p. 42.
13. "For, methinks, the *understanding* is not much unlike a closet [dark room] wholly shut from light, with only some little opening left, to let in external visible resemblances, or ideas of things without" in John Locke, *An Essay Concerning Human Understanding* [1689], London, Penguin Books, 1997, p. 158. The model—extended in the spirit of Descartes who had made it an image of the eye—is that of the *camera obscura*.
14. There can be no nature without technique, at the very least because nature only exists for the human being in a world where there are human beings, i.e., a technologized world. Nature designates a certain state of technology—and conversely technology is not the prerogative of the human being (which we will probably agree upon more easily). All perception is a tooling.
15. Philippe-Alain Michaud, "Étienne-Jules Marey et la poétique des mobiles" [1996], *Sketches: Histoire de l'art, cinéma*, Paris, Kargo and L'Éclat, 2009, p. 55.
16. Dominique Chateau, *Philosophie d'un art moderne: Le cinéma*, Paris, L'Harmattan, 2009, p. 65.
17. Érik Bullot, "Animal flicker", *Trafic*, no. 109, Spring 2019, p. 5.
18. Paolo Gioli, in S. Toffetti and A. Licciardello (eds), *Un cinema dell'impronta*, Rome, Centro Sperimentale di Cinematografia, 2009, http://cherrykino.blogspot.com/2011/03/paolo-gioli-artisan-filmmaker-free.html (excerpts).
19. Michel Thévoz, *Dubuffet*, Geneva, Skira, 1986, p. 104.
20. Ibid., p. 105.
21. Érik Bullot, "Empreinte et cinéma", *Cinémathèque*, no. 13, 1998, p. 116.
22. Yann Beauvais, *Poussière d'image*, Paris, Paris Expérimental, 1998, p. 82.
23. See for example, Bruce R. McPherson (ed.), *Essential Brakhage: Selected Writings on Film by Stan Brakhage*, New York, Documentext, 2001, pp. 28–29.
24. J. W. von Goethe, *Theory of Colours* [1810], trans. Charles L. Eastlake, Cambridge, Mass., MIT Press, 1970, p. 40. For more details,

see Jonathan Crary, *Techniques of the Observer: Vision and Modernity in the Nineteenth Century*, Cambridge, Mass., MIT Press, 1990.

25. Julien Gracq, *Reading Writing* [1980], trans. Jeanine Herman, New York, Turtle Point Press, 2006, p. 279.
26. André Bazin, "Ontology of the Photographic Image" [1945], *What is Cinema? Vol. 1*, trans. Hugh Gray, Berkeley, University of California Press, 2005, p. 15.
27. "This documentary aspect, called upon (by photography and cinema) to safeguard *forms* from the ravages of time, has in fact introduced into them time itself, that is to say a kind of biological category of nature: it has made them age or seem "past" [*les a fait vieillir ou "passer"*], something that neither the fragile manuscript, the delicate painting nor the uncertain memory had produced [...]. The mechanical image renders out of date that which it records [*périme ce qu'il enregistre*]—it introduces it to time, to the time of consumption in a solicitation or suggestion of an imitation of short duration" in Jean Louis Schefer, *Du monde et du mouvement des images*, Paris, Cahiers du cinéma, 1997, p. 10.
28. This is the continuation of the ecology of restoration initiated in the 1830s by the reintroduction of the Capercaillie in Scotland. See D. Jorgensen, "Reintroduction and De-Extinction", *Bioscience*, no. 63, 2013, pp. 719–720.
29. Deleuze had already written this about becoming-animal: "The becoming-animal of the human being is real, even if the animal the human being becomes is not" in Gilles Deleuze and Félix Guattari, *A Thousand Plateaus. Capitalism and Schizophrenia* [1980], trans. Brian Massumi, Minneapolis, University of Minnesota Press, 1987, p. 238.

CHAPTER 5

The Aesthetic of Contact in Contemporary Photochemical Film Practice

Kim Knowles

Since the turn of the millennium, the predominance of digital technology has given rise to an increased sensitivity to celluloid film as a tangible medium—its unique capacity to record the indexical traces of the filmmaker's hand, as well as other physical encounters, distinguishes it from these newer forms of image making and creates a framework for rethinking film in a digital era.[1] Although proclamations about the imminent disappearance of photochemical film were frequent during the early years of technological transition, a sustained wave of interest in 16 mm production has, for the moment at least, guaranteed the medium a second life on the margins of mainstream film production. It now flourishes in the field of experimental and artists' film, offering a playground for exploration and innovation facilitated by the increased availability of outmoded equipment and a culture of DIY chemistry and skills sharing. From within this context of artistic autonomy emerges a new aesthetic of materialism

K. Knowles (✉)
Aberystwyth University, Aberystwyth, UK
e-mail: kik2@aber.ac.uk

E. Della Noce and L. Murari (eds.), *Expanded Nature*, Experimental Film and Artists' Moving Image,
https://doi.org/10.1007/978-3-031-70729-2_5

that both draws and expands on its historical roots. It looks backwards to project forwards, cognizant of its liminal cultural status that plays out in the methods employed and the forms of vision to which they give rise.

This chapter maps out directions for a renewed approach to materialist film that locates contemporary engagements with the physicality of celluloid within emerging theories of new materialism. I argue that this field opens up productive avenues for exploring current trends in photochemical film practice where alternative forms of vision are a key concern. Central to this investigation is the growing contemporary interest in film's ability to reveal haptic forms of knowledge that transcend the purely optical and negotiate the space between the visible and the invisible. Focusing primarily on bodily and earthly engagements, I seek to show how contemporary materialist film is located within discourses on the environment and ethical modes of being in the world.

The increased awareness of celluloid film as a living (and dying) body, thrown into relief by its digital counterpart, creates a context for revised formulations about its representational capacities. That film has a certain material "life force" is what ties it specifically to a wave of theoretical work that attempts to understand the world as an interconnected web of energies, "a multitude of interlocking systems and forces", where animate and inanimate matter, the human and the non-human, exist in a constant non-hierarchical dialogue.[2] Loosely assembled under the umbrella term "new materialism", these theories focus attention on the concept of agency, challenging the commonly held assumption that objects, things, and matter are passively acted upon by the human subject. By arguing that the physical world is constituted through a process of continuous change that plays out both within and beyond the realms of human intervention, new materialism forces us to reconsider reality as fundamentally unstable and thus unknowable to us through vision alone. This is important to contemporary reflections on and experiments with film as a material substance that seek to uncover its hidden agential capacities and its ability to dialogue with other tangible matter. In the works discussed in this chapter, process rises to the surface as evidence of the often invisible chemical encounters that take place on the filmstrip, allowing the spectator to move between the layers of representation, from an immersion in the image to a contemplation of the materials and gestures involved in its coming into being. It is through the performativity of process, therefore, that new forms of experience and modes of contact with the world are translated and celebrated.

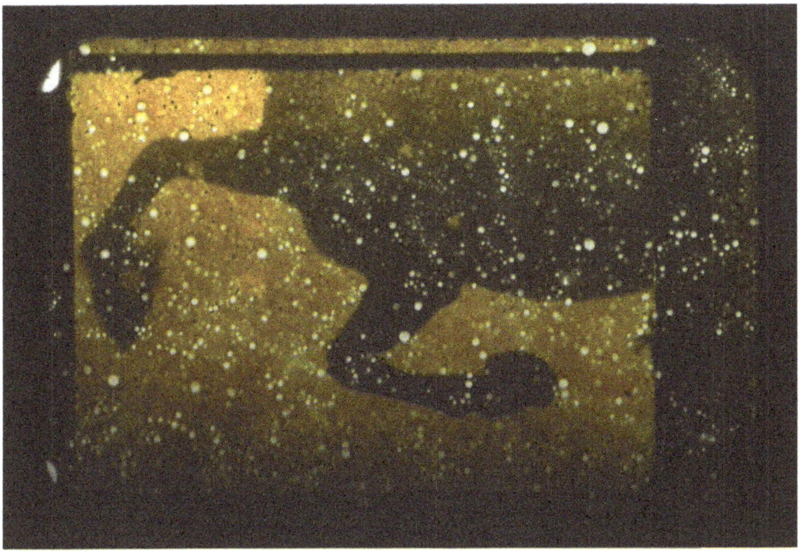

Fig. 5.1 Esther Urlus, *Konrad & Kurfurst* (2013–2014) © All rights reserved by the artist/Courtesy of Light Cone

In Esther Urlus's *Konrad & Kurfurst* (2013–2014), the fragmented image of a horse shifts in and out of view, struggling against a bubbling viscous background. The figurative image hovers in a space of indeterminacy, revealing itself fleetingly, only to be dragged back into the sticky stuff of matter (Fig. 5.1).

Urlus describes the film as a fictional re-enactment of a five-minute incident in the 1936 Berlin Olympic Games, in which Konrad Freiherr von Wangenheim was thrown violently from his horse Kurfurst, but nonetheless went on to win a gold medal. However, the home-brewed film emulsion that the artist fabricated after consulting numerous technical manuals of early film and photography ultimately constitutes the main focus of the work, functioning, in the filmmaker's words, "as a fragile metaphor for the heroism of Konrad and his horse Kurfurst". Urlus is also clearly referring to the fragile cultural status of celluloid, and the pressing need for artists to explore alternatives to the dwindling commercial manufacture of film stock. Konrad's fall, and ultimate unforeseen triumph, could be read as a reflection on the narrative of film itself and its potential surmounting of the challenges of obsolescence. The struggle is an allegory—the fight to stay alive against all odds—and from this perspective the foregrounding of the material is multi-layered in all senses of the word.

The emulsion is subject to constant change through time, eluding a stable perceptual ground against which the figure might be distinguished and fixed. Movement animates the dried bubbles of emulsion and in the moment of projection appears to turn them back into liquid, reversing time in the process of moving the film forwards. Through a series of chemical interactions, then, attention is shifted to the surface of the filmstrip and its material presence/present. By bringing the support—the handmade film emulsion—into the visual field Urlus arguably stages what Jane Bennett describes as the "vibrancy of matter",[3] and illustrates Barbara Bolt's assertion that "art is a co-collaboration [...] and matter as much as the human has responsibility for the emergence of art. In other words, 'matter has agency'."[4] What is significant about *Konrad & Kurfurst* in this context is the insistence on process as the site of meaning. If the film tells a story, it is one that repeatedly brings us back to material relations. The shifting effervescence of emulsion on the surface of the film frustrates vision in order to call attention to material instabilities and contingencies and asserts its own agency beyond the controlling hand of the artist.

In the spirit of experimentation and re-discovery, current DIY film culture opens up to chance and accidental occurrences, taking film chemistry back to its primitive origins, not as a nostalgic exercise of longing for the past, but as a playfully progressive act that challenges traditional values of perfection and stability and sets out possible pathways for future explorations of analogue film. As Urlus states in the booklet that emerged from her research into handmade film processes: "For me film is a lot more than the representation of what a camera has recorded—exposure and development. It's important for me that the emulsion has its own texture and can respond to chemicals differently than expected."[5] The filmmaker thus privileges a process in which her own artistic agency gives way to the eventualities of these material encounters. Texture, in the form of material presence, is the means by which the film communicates, as it represents the meeting point of the chemical transactions and transformations—a process that, while invisible to the viewer, is nonetheless contained in the images they see and also sense.

The recent wave of handmade film practice demonstrates a deep investment in exploring how film responds to matter. West Coast film artist Jennifer West, for example, marinades film in food substances before submitting it to numerous material interventions that are signalled in the playfully irreverent titles such as *Whatever Film: 16 mm film leader soaked in lots of coffee and espresso, taken on a power walk, rubbed with sweat and inscribed with the word "whatever" written in purple metallic eyeliner* (2007). As West's title suggests, the human body also becomes a site of material exchange. In works such as Thorsten Fleisch's *Bloodlust* (1998) and Vicky Smith's *sobbingspittingscratching* (2010), bodily waste is applied to the surface of the film, reanimating liquid secretions and providing a means to render visible (and sensible) the internal flux and flow normally hidden by the reversible layer of skin that simultaneously connects and separates us from both ourselves and the world.[6] The "skin" of the film therefore acts as a mediator, bringing the two bodies into unison, with each enacting physical change upon the other. Like the buried films I will go on to discuss, this form of material engagement draws on the power of the indexical trace to extend the representational spectrum to phenomena that traditionally exist outside our visual field.

These body fluid films, as I have argued elsewhere, shed light on the general status of waste in contemporary society.[7] Rejected from the social sphere and relegated to geographical peripheries, waste is the invisible residual, the abject other side of capitalist consumption from which we must separate ourselves in order to maintain a sense of bodily unity. As John Scanlan points out, waste reminds us too much of our own mortality, and our ultimately perishable nature. "[T]he creation of garbage", he argues, "results from a more or less imperceptible contest between life and death—because death constitutes the human return to matter".[8] New materialist theory takes important steps towards understanding the matter of waste from a productive perspective, delineating this liminal space between life and death (the space that film itself occupies) as replete with a vibrant force that emanates from and sparks between discarded objects wrenched from their original context. Describing an encounter with a littering of disparate items on a street in Baltimore—a plastic work glove, a chunk of oak pollen, a dead rat, a plastic bottle cap, and a stick of wood—Bennett argues that the affective draw of such a scene demonstrates how "thing power [can rise] from a pile of trash".[9]

In fleshing out this idea of "thing power", Bennett draws on Louis Althusser's notion of "aleatory materialism" as a means to reconceptualize our material world as a state of flux rather than consisting of stable entities and to support the idea that "so-called inanimate things have a life, that deep within is an inexplicable vitality or energy".[10] Althusser's theory is an attempt to develop a framework for thinking about the positioning of the historical subject in relation to systems of power and state control, but its engagement with materialist philosophy as the foundation for understanding structures of being allows it a wide-ranging application in contemporary considerations of matter. Important to the present discussion of physical interactions in film is Althusser's assertion that:

> Every encounter is aleatory in its effects, in that nothing in the elements of the encounter prefigures, before the actual encounter, the contours and determinations of the being that will emerge from it ... no determination of these elements can be assigned, except by *working backwards* from the result to its becoming, in its retroaction.[11]

This has very particular implications for a new conception of materialist aesthetics that does not presuppose the existence of material specificities but rather redirects attention towards film as an alchemical process with indeterminate outcomes. The practices discussed here foreground this working backwards, inscribing temporality and change into the experience of the work. The fixed notion of what film *is* no longer holds the centre of interest; it is rather the question of how film *becomes*, in relation to other material phenomena, that animates the field of celluloid practice and propels it into new areas of visual exploration. As Pip Chodorov states, "we don't work with 'images,' but with organic, physical material that comes from the earth: salts, silvers, minerals".[12] The material substrate of film is therefore always already a complex combination of elements—an amalgam of vibrant matter whose chance constellations lead to a continually changing granular makeup. Chodorov's connection of film with earthly matter 'touches' on the nature of film as a substance that holds within it the possibility for opening up new modes of vision, where the figurative image collapses into its material makeup. It is here that we find the most fertile terrain for unravelling the relevance of Althusser's concept of the encounter, and for exploring the relationship between the ecological concerns of new materialism and the increasing interest in the environment within contemporary materialist film practice.

"For the current cinematic avant-garde", states Tess Takahashi, "it is less film's ability to produce recognizable iconic images of the natural world indexically that is emphasized, than its ability to physically record the influence of the material world on its celluloid body".[13] As Takahashi's overview illustrates, cameraless engagements with the environment have become a mainstay of experimental film practice since the rise of digital technology at the turn of the millennium, emerging partly from the drive to differentiate the physicality of film from the apparent intangibility of digital. In this sense, the alignment of film with the natural world might be seen as a defensive strategy on the part of film practitioners—a way of carving out a position of autonomy for a precarious medium on a slippery slope to obsolescence. It is impossible to avoid such associations, contentious and divisive as the film–digital debate tends to be, but this should not preclude a serious contemplation of these engagements since the new forms of vision and tactile knowledge to which they give rise transcend any simplistic claims about nostalgia and media fetishism. In particular, the new aesthetic of contact that arises from these contemporary collaborations with the environment offers a more multifaceted approach to medium specificity, where attention is displaced from the film itself to the processes and encounters involved in its material transformation.

An early example is David Gatten's *What the Water Said, nos 1–3* (1997), in which several unspooled rolls of film were placed in a crab trap and thrown into the sea. The abstract images in/on the film are nothing other than a register of a dialogue between the film material and the changing conditions of the sea depending on weather and tidal movement. Using single-perforation sound stock, the filmmaker was able to create a corresponding sonic record of the encounter, which recreates for the viewer the experience of being submerged in water.[14] In his discussion of the film, Scott MacDonald points to a crucial shift in perspective away from content and towards process, arguing that "[w]hatever the effects of Gatten's flinging the crab trap into the sea would have created", the act itself reveals "a faith in the possibility of collaborating with the environment in a more direct way".[15] The image is therefore secondary to the gesture, or, rather, it exists primarily as a record—a trace—of a physical encounter that is both durational and, importantly, inaccessible in its complexity to both the filmmaker and the viewer. The film can be understood as working on two interrelated levels: on the one hand it forges a deeper material (embodied) engagement with what it represents, and on the other it testifies to an experience (a process) that cannot easily be turned into an image, cannot be owned, penetrated, and thus commodified—in this, the film both alludes and *e*ludes. The other side of the film's refusal of perceptual clarity is its embrace of the unknown—that which escapes vision but not representation.

Another filmmaker to have interrogated the layers of representation in relation to the environment is Emmanuel Lefrant, whose *Parties Visible et Invisible d'Un Ensemble Sous Tension* (2009) is directly inspired by Maurice Merleau-Ponty's 1964 text *Le Visible et l'invisible (The Visible and the Invisible).* Lefrant uses the process of making the film as a way of directly illustrating Merleau-Ponty's theory of perception, in which the body and the world, subjective experience and objective existence, are interwoven. This new conception of the body that emerges in the philosopher's later work is an attempt to articulate the relationship between "what *seeing* is, and what *thing* or *world* is".[16] Vision, for Merleau-Ponty, is inseparable from touch: "The look ... envelops, palpates, espouses the visible things. As though it were in a relation of pre-established harmony with them, as though it knew them before knowing them."[17] In *Parties Visible et Invisible* material processes are employed to translate—through touch—an earthly experience that demonstrates the reversibility of perception. On Super 8 film Lefrant shot a single static image of a landscape seen from a window: a bush in the distance, with the horizon line between land and sky splitting the image in two (Fig. 5.2).

Several strips of processed but unexposed Super 8 and 16 mm film were then buried in the ground at the site of shooting and unearthed several weeks later—enough time for the biochemical breakdown of the emulsion to be visible on the surface of the filmstrip. The abstract (buried) and figurative (filmed) images are sandwiched together and interlaced through the process of bi-packing—an optical printing procedure that superimposes the details of one image onto another. The end result is a startling animation of the figurative image by the colourful degradation and decomposition of the buried strip. The transposing of these qualities—the invisible (underground) onto the visible (overground)—is emphasized by the visual impression of flipping that caresses the eye as a tactile flicker. Lefrant refers to the film as "a bipolar world, where the invisible takes shape within the visible, where the first dissolves into the second and vice-versa".[18] As the different sections of the film become interlaced, as subjective becomes objective, inside becomes outside, and invisible becomes visible.

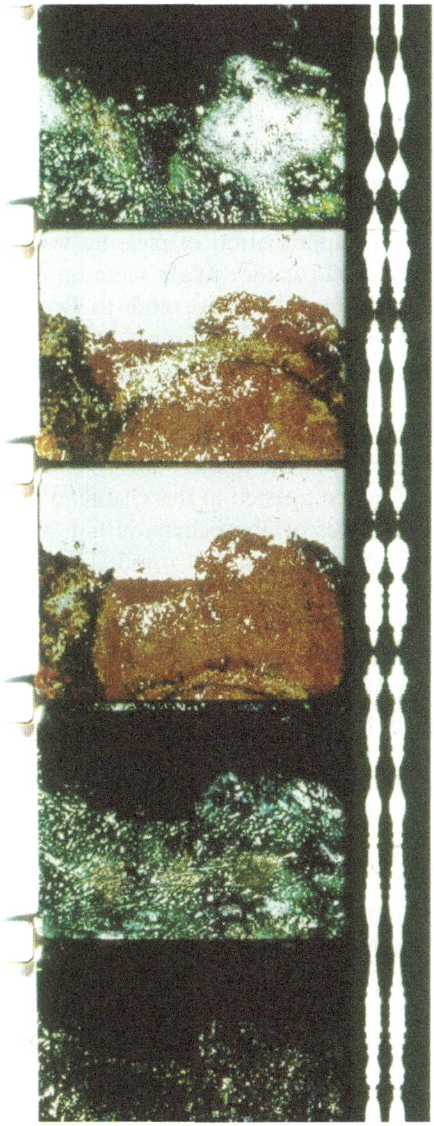

Fig. 5.2 Emmanuel Lefrant, *Parties Visible et Invisible d'Un Ensemble Sous Tension* (2009) © All rights reserved by the artist/Courtesy of Light Cone

Running under the surface of Lefrant's film, as well as his more recent works such as *Le Pays Dévasté* (2015) and *I don't think I can see an Island* (2016), is the theme of environmental catastrophe in the age of the Anthropocene. The fragility of the filmstrip and its susceptibility to physical intervention finds its analogy in the images of barren landscapes ablaze with energy and urgency. There is something deeply apocalyptic about these films that undermines or at the very least problematizes a straightforward appreciation of pleasing visual effects and material specificities. As beautiful as they might seem on one level, it is impossible not to read the fiery explosions of colour in *Parties Visible et Invisible* as a deterioration not just of the filmstrip, but of the earth itself, and the destruction of the natural world. Like Gatten, Lefrant draws on the potential of film to communicate forces that elude traditional methods of representation, where the gap between visibility and invisibility is narrowed through the process of tactile translation.

I have suggested in this chapter a number of ways to consider contemporary uses of photochemical film as developing new modes of physical engagement and haptic vision. The discussion has focused particularly on artisanal, handmade, and cameraless film, which although by no means representative of all current analogue practice, nonetheless constitutes a significant wave of activity emerging partly as a response to the medium's now marginal cultural position. Working by hand is both an economic and an aesthetic decision that allows the artist an increased sense of creative freedom and autonomy from industrial processes. However, it also, as we have seen, facilitates direct engagement with other material phenomena such as chemicals, body fluids, soil, and water—an amalgam of 'stuff' that is ripe, in the hands of the experimental filmmaker, for tactile investigation and re-appropriation, resonant with vibrant energies that speak to and through the celluloid substrate. Crucial here is the intervention of new materialism in staking out modes of enquiry into how we experience and interact with the world, giving rise to new subjectivities that shatter the human/non-human hierarchy. As much scholarship in this area points out, in an age of environmental decline and the worrying cycles of production and disposal that give rise to increasing quantities of material waste, appropriate ethical positions are required to reinstate a sense of responsibility. Employing the concepts put forward by Jane Bennett, I have argued that the foregrounding of materials and materiality in contemporary works demonstrates new forms of embodied knowledge through the sensuous encounter with matter. We might, in this sense, connect these gestures with a form of political thinking that starts with a reconsideration of how we view and interact with the physical world.

Notes

1. A slightly longer version of this article was first published as "(Re)visioning Celluloid: Aesthetics of Contact in Materialist Film," in Martine Beugnet, Allan Cameron, Arild Fetveit (eds), *Indefinite Visions: Cinema and the Attractions of Uncertainty*, Edinburgh, Edinburgh University Press, 2017, pp. 257–272.
2. Diana Coole and Samantha Frost, "Introducing the New Materialisms", *in* Diana Coole and Samantha Frost (eds) *New Materialisms: Ontology, Agency, and Politics*, Durham, Duke University Press, 2010, p. 9.
3. Jane Bennett, *Vibrant Matter: A Political Ecology of Things*, Durham, Duke University Press, 2010.
4. Barbara Bolt, "Toward a 'New Materialism' Through the Arts", in Barbara Bolt and Estelle Barrett (eds), *Carnal Knowledge: Towards a 'New Materialism' through the Arts*, London and New York, I.B. Tauris, 2013, p. 6.
5. Esther Urlus, *Re:Inventing the Pioneers: Film Experiments on Handmade Silver Gelatin Emulsion and Color Methods*, Rotterdam, Self-published, 2016, p. 2.
6. On the subject of film as skin, see Jennifer Barker, *The Tactile Eye: Touch and the Cinematic Experience*, Berkeley, University of California Press, 2009.
7. Kim Knowles, "Blood, sweat and tears: Bodily inscriptions in contemporary experimental film", *NECSUS: European Journal of Media Studies,* vol. 4, 2013, pp. 447–463.
8. Scanlan, John, *On Garbage*, London, Reaktion Books, 2005, p. 9.
9. Jane Bennett, *Vibrant Matter, op. cit*., p. 6.
10. Ibid., p. 18.
11. Louis Althusser, *Philosophy of the Encounter: Later Writings, 1978–87*, trans. G.M. Goshgarian, London, Verso, 2006, p. 193.
12. Pip Chodorov, "The Artist-Run Film Labs", *Millennium Film Journal,* no. 60, Autumn 2014, p. 36.
13. Tess Takahashi, "After the Death of Film: Writing the Natural World in the Digital Age", *Visible Language,* vol. 42, no. 1, 2008, p. 49.

14. Scott MacDonald, *The Garden in the Machine: A Field Guide to Independent Films about Place*, Berkeley, University of California Press, 2001, p. 374.
15. *Ibid.*, p. 375.
16. Maurice Merleau-Ponty, *The Visible and the Invisible*, Evanston, Northwest University Press, 1968, p. 3.
17. Ibid., p. 133.
18. http://lightcone.org/en/film-5833-parties-visible-et-invisible-d-un-ensemble-sous-tension [accessed 28 September 2015].

CHAPTER 6

Booming Harmonics: Feminist Ecologies of Process Cinema

Philip Hoffman and Janine Marchessault

This chapter will explore recent practices of 'Process Cinema'[1] that are informed by understandings of plant and animal ecologies, feminist phenomenologies, and entangled temporalities of the planetary. Examining recent films that are deeply situated through techniques of animation and DIY filmmaking, as well as traditions of women's handiwork such as quilting, stitching, and beading, the chapter reflects upon recent forms of 'ecological thinking' in feminist experimental films. Such works include living systems outside the frame grounded in actions of material making. But first we start with history.

P. Hoffman (✉) · J. Marchessault
York University, Toronto, Canada
e-mail: hoffmanphilm@gmail.com

J. Marchessault
e-mail: jmarches@yorku.ca

E. Della Noce and L. Murari (eds.), *Expanded Nature*, Experimental Film and Artists' Moving Image,
https://doi.org/10.1007/978-3-031-70729-2_6

Botany

Anna Atkins (1799–1871), British amateur botanist, was the first person to publish a book of photographs with her beautiful *British Algae: Cyanotype Impressions.* The impressions were all developed by hand between 1843 and 1853. As Larry Schaaf describes:

> Atkins positioned her dried seaweed on the coated paper and placed this under glass in the sunlight. Within minutes, solar energy had worked its sleight of hand and a trace image became visible. Plunged into plain water, the affected iron compounds formed the familiar pigment Prussian blue. Acting the role of photographic "negative," the specimen was used to make additional copies of that plate, each a faithful translation of the original plant.[2]

The images in the *British Algae: Cyanotype Impressions* were each produced by hand (her hands and with the help of solar energy's "sleight of hand"), which is why the dozen copies of the book took more than a decade to produce. These represent the first applications of photography for scientific study using the discoveries of two of Atkins' friends. The first came to her from amateur botanist William Henry Fox Talbot, who recommended that she use photography to study plants. After testing her silver-salt process for her book, she ended up applying the "cameraless" invention of another friend, John Herschel, "whose cyanotype—a blue printing process—made use of the sensitivity of ferric salts to light".[3] Atkins' brilliance was to use cyanotypes in a series of books in which she used photography as scientific illustration and that she also created as unique gifts for her circle of friends.

Botany was the only professional field that women had access to in the 18th- and 19th-century Anglo-American world.[4] Atkins' father John George Children (Director of the British Museum of Natural History) had greatly supported her research and her career as a botanical illustrator. After the death of her father, her close friend, botanist Anna Dixon, whom Children had raised alongside Atkins, worked with her to produce several more books in his commemoration. To the study of algae she added ferns, lace, and feathers whose patterns were not only complex but aesthetically pleasing. Indeed, as Annemarie Iker has pointed out, the later works of Atkins focus more on aesthetics and abstraction, and less on scientific veracity.[5] Larry Schaaf wrote the definitive essay on Atkins' contributions, stating that she had all but disappeared from history, since when gifting her books to friends, she often signed them "A.A.", which had been interpreted as "Anonymous Amateur". Therefore acknowledgment of her work was lost for many years. It is thanks to Schaaf and other scholars that Atkins' contributions have now been recognized.[6]

Atkins' extraordinary experiments with plants help us to introduce the idea of ecological thinking. Joyce Wieland's mixed media, films and quilts, provide us with powerful examples of ecological consciousness inspired by the 1960s civil rights movements tied to feminist and Black activist struggles, as well as the emerging discourses of environmental politics in North America. We can define ecological thinking as a recognition that takes into account the interconnectedness of all life forms, and depends upon the profound connection between inside and outside spaces.

In the films discussed in this chapter, the integration of the outside world expressed through the incorporation of natural elements (the atmospheric and elemental, flowers, plants, animals, and the cosmos) is mediated through touch and women's handiwork. As we explore in the chapter's last section, devoted to the films of the Odeimin Runners Club, this notion of the interdependencies of life forms on this planet is profoundly influenced by Indigenous ways of knowing.

Ecology

True Patriot Love, by Joyce Wieland, was the title of the National Gallery of Canada's first solo exhibition of a living woman artist in 1971. As indicated in her artist statement, the central theme of Wieland's exhibition and exhibition catalogue is "ecology". The exhibition catalogue was produced as a singular artbook created by appropriating a 1964 government publication devoted to Arctic flora. The book is striking in terms of its collage aesthetic and material dynamics. In her creation of an ecology, she explores a heterogeneous landscape of diverse images, forms of writing, newspaper clippings, a film script devoted to the Canadian painter Tom Thompson, drawings, photographs, poems, and songs in different languages (English, French, Inuktitut, and Gaelic). Wieland's artistic intervention of the collaged government document resembles a scrapbook that is fastened together by women's handiwork—pages are sown together, recipes for creating fabric and film dyes are included and, in reference to those 19th-century women botanists, flowers are pressed into the body of the book (Fig. 6.1).

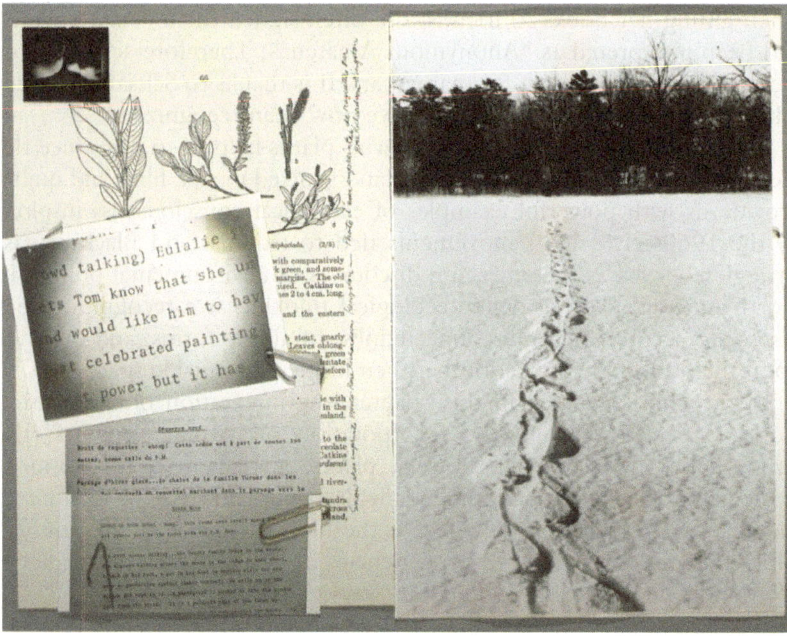

Fig. 6.1 Joyce Wieland, Catalogue, *True Patriot Love* (1971) © All rights reserved by the artist/Courtesy of Light Cone

The book itself becomes a living document of her handmade art and her relation to Canada as a political territory and an environment. In keeping with the idea of ecology as interconnectedness, an opening to the outside—of bringing the outside in—Johanne Sloan has pointed out that Wieland's exhibition expanded the institution to include everyday realities:

> Wieland treated the gallery as a social space, that is only provisionally isolated from a larger world of politics and everyday life. ... When one newspaper critic sneered that 'Joyce the housewife empties out her attic and her barnyard and fills... the National Gallery with pillows and quilts', he was actually close to understanding the artist's intent. She was urging the nation's most prestigious art institution to enter into a dialogue with everyday life (even that of housewives!) and to engage with the important political issues of the day.[7]

The concept of ecology that Wieland was working with no doubt came from her connection with the New Left, as Sloan carefully traces, and also through a connection to pop art and its emphasis on media ecology. Wieland has created ecological frameworks that expand traditional forms of film by bringing in traditions specifically belonging to women, from embroidery and quilting to fabric dying and tinting. In *Water Sark* (1964–65) Wieland films kitchen objects (crockery, rubber gloves, a teapot) as well as her own body to discover and define a feminine space; fabric dyes are applied directly to footage, and quilting needles are used to create perforations in *Hand-Tinting* (1967–68); and small domestic animals such as cats and hamsters appear in her fairytale *Rat Life and Diet in North America* (1968). Her use of media materiality allowed her to expand the film frame and move between cultural and performative spheres. It is the 'home-made' aspects of filmmaking that are emphasized, just as some of her plastic wall hangings are made to look like film strips ('*Stuffed Movie*,' '*Home Movie*,' '*War and Peace 8 mm Home Movie*' (1966)). Many of these works across media including her paintings reference the handicraft to recall craft art as an empowering, collective (even sacred) activity for women. The politicization of women's labor including craft and handicraft was intrinsic to the burgeoning feminist art cultures in New York that she was connected to before moving back to Canada. Certainly, Wieland was a leader in the emerging field of feminist art in the late 1960s and beyond. Her concept of an expanded image, of an ecology of images, set the stage for the next generation of ecologically conscious filmmakers.

When It Was Blue

Jennifer Reeves' work arises from the materialist practices that Wieland helped to initiate. Her foray into working with hand-processed film which she tinted and toned occurred when she attended the Independent Imaging Retreat in Mount Forest in 1997 (she attended again in 2011). Colloquially known as "Film Farm", it has been running for over 25 years and was established by Marian McMahon and Philip Hoffman as a seven-day retreat, combining feminist pedagogy with experimental filmmaking, committed to supporting feminist and queer filmmakers. The workshop initiated a DIY anti-industrial and collaborative way of working with film, certainly inspired by the work of Wieland and the handmade process approach to working with film through material location-based hand-processing of celluloid. Other artisanal approaches at the workshop include scratching, painting, decaying, tinting, toning, solarization, optical printing, and more recently working with native plants and seasonal flowers used in film processing, tinting, and phytogram-making. The latter was initiated by a workshop given by visiting artist Karel Doing in 2018, whose teachings carried through into future Film Farm workshops. Many of the participants of these 'eco-film' workshops, such as Franci Duran, Dawn George, Madi Piller, Emily Pelstring, Ramey Newell, and Tara Khalili, integrate plant-based processes into their film and art making.[8] The workshop now exclusively supports non- or low-toxic approaches to developing and manipulating film.

Jennifer Reeves' short hand-processed film *We are going home* (1998) is a beautiful surrealistic film work that was made in the early stages of Film Farm's development. The film experiments with numerous DIY cinematic approaches such as in-camera superimposition, optical printing, tinting, and split toning. It is remarkable for its masterful control of color and texture to create natural landscapes which host queer erotic fantasies and uncanny reveries. Reeves further developed techniques of split toning in 2011, when she returned to the Film Farm and made *Strawberries in the Summertime* (2013), a mysterious view of the natural world from the perspective of her child.

Reeves has become one of the most important film artists working with celluloid through optical and hand-painted processes in North America. Her astounding 2008 film projection performance *When It Was Blue*—a collaboration with Icelandic bass player and composer Skúli Sverrisson—was presented to create a superimposed dual projection performance. The work on 16 mm film is a breathless one-hour experience that provides a stream of hand-painted superimposed optically printed images of ecosystems around the planet from New Zealand and Iceland to the Americas with a focus on flora and fauna. The film has been described as a "rapturous homage to the endangered beauty of our blue planet".[9] But as Michael Sicinski remarks:

> *When It Was Blue* is a far more complex, more ambivalent work of art. Reeves' relationship to the natural world, to the world at large, and to cinematic meaning-making conveys a trembling, polyvalent sense of the communicative act. It is a dialectic that cannot resolve, that mere montage will not suffice to explicate. It is visceral, not rhetorical. And it needs

> multiple images hitting the screen, complicating one another at the same time, at all times. It is one singer singing with two mouths, live.[10]

Sicinski's essay on Reeves' performance is a kind of apology for having completely dismissed her films prior to this. However, in this review, he apologizes for his earlier misstep. This kind of purposeful omission from the canons of experimental film is something that Wieland and other women such as Carolee Schneeman, Barbara Hammer, Barbara Sternberg, Frédérique Devaux, and many others have long suffered. Sicinski notes that Reeves has created a sophisticated dialectical framework of objective (male, Western, masterful) and subjective (hand-painted and intimate) world images which collide and contradict each other, standing both outside and inside at the same time through live superimpositions with analogue projectors; a dialectic which leads not to synthesis but to open-ended reflections on a world in motion and in the process of disappearing. There is a sense that Reeves is capturing the images and painting with them in order to bring us in as accomplices in imagining new worlds. The film performance is structured in four parts—representing the seasons and the traveler's (the artist) journey across different spaces. There are more flora and fauna in the film than people—whom we see occasionally present in images, often as tourists or spectators of natural wonders such as the ocean, seen both from a distance and in close up from a boat. These diverging sequences represent different temporalities that are superimposed. The film opens with the ocean and the moon—two things that are blue. Across all four parts, a small plane circles, at one time framed by the moon, to survey the land below. It quickly moves us to boat tankers and an oil rig. The stunning hand-painted film frames dominate the film and function almost as stained-glass windows through which to view the fleeting worlds, weather events (storms, tornados, fires), plants and animals, oceans and forests. Vibrant red, orange, green, and of course blue are painted into the light sections of the images (Fig. 6.2).

Alternatively, we are confronted with thousands of paintings (both still and moving) which appear layered, sometimes abstract, completely covering the images that are moving past our eyes in flash impressions. The effect is breathtaking. While stained-glass windows in their medieval incarnations were intended to lift the beholder's consciousness into the spiritual realm—Reeves' cinematic windows bridge our consciousness to the outside/inside of her phenomenal perceptions. The painting is an action forward, each frame sculpted by diverse light patterns captured

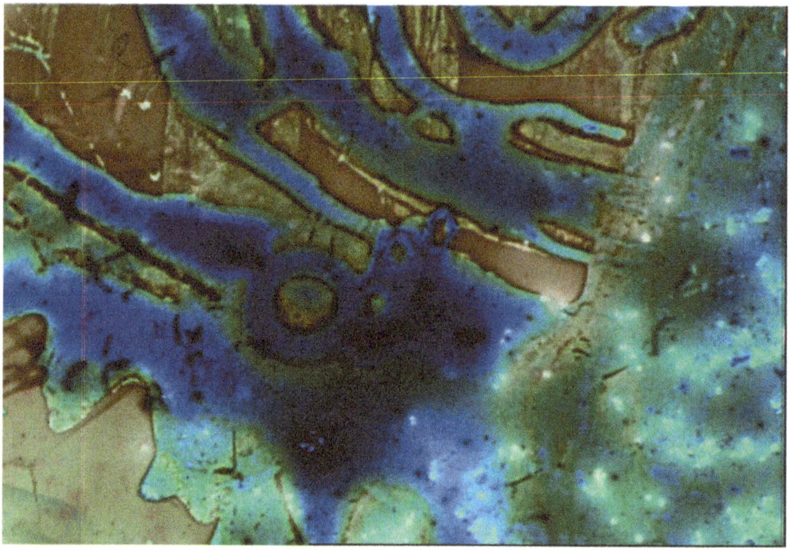

Fig. 6.2 Jennifer Reeves, *When It Was Blue* (2008) © All rights reserved by the artist/Courtesy of Light Cone

through images of nature, destruction, and climate change—ruins taken over by foliage, oil spills destroying animal life, massive pieces of glaciers melting away—all reflected and mirrored by the actions of the painted frames. The experience of light here is tactile; haptic maneuvers which stream past us. The film performance is based in contingency and/of the present moment of being here together.

When It Was Blue was made just as the term Anthropocene was coming into common usage, signaling a new approach to nature and to history defined by the complete entanglement of human history and natural history—the slow history of repetitive cycles has been profoundly disrupted by the processes of capitalist over-production and resource extraction. Historian Depesh Chakrabarty argues that with human history and geological history so completely intertwined, a new conceptual framework is needed that combines earth and worlds. Humans are part of "a force" that combines three histories—the history of planetary systems, the history of life on the planet, and the history of capitalism. It is critical to understand that these histories operate at different scales and speeds.

Chakrabarty argues that artists will necessarily serve as the creators for a new conceptual framework to help us recognize and reflect on these new entanglements.[11] This is precisely what Reeve's film does—it gives us a glimpse into these complex entanglements and temporalities of earth and worlds, the now-time of the performance and the deep time of the natural world without proposing easy solutions or falling back on simple nostalgia or ideology critique. We are beyond that.

While Sicinski points out that Reeves' work is aligned with an earlier generation of mythopoetic filmmakers such as Stan Brakhage (painting on film), Brenda Longfellow has argued that Reeves' representational strategies are significantly different from Brakhage's "adventures" in perception. Not only does Reeves sculpt a rich acoustic soundscape for the film (contra Brakhage's silent films), but her film actively counters the perceptual abstraction characteristic of Brakhage's cinema. In her contrapuntal sound–image juxtapositions, Reeves crafts a "complex and mediated social commentary" on the ecological discourse surrounding the "blue planet".[12] We would argue with Longfellow that Reeves' "eco-impulse"[13] is building upon a feminist tradition of expanding the frame to include the outside world which is beyond her/our control, our grasp and understanding and yet, as we see in the film, is often reduced to instrumental extraction—practices which Reeves decimates with her brush. Those feminist traditions are not only found in the media works of Joyce Wieland but also in Carolee Schneemann's radical performances to "explode the canvas".[14] *Fuses* (1967) in particular is the best example of this explosion, layering intimate sexual footage of herself with quotidian images of her surroundings, all 'fused' with paint, fire, and acid, and other found objects to scratch and create a cine-collage. Schneemann's diverse works have helped to inaugurate practices of painting on film and also new kinds of embodied performances incorporating diverse media, objects, and celluloid.

Reeves is working through the lens of an eco-feminist epistemology which stands quite apart from the mythopoetic tradition. The connection between women, non-human animals, and forms of patriarchal domination that Reeves explored in her earlier films and especially in her work which explores agoraphobia and 9/11, *The Time We Killed* (2004), is very much present in *When It Was Blue*. This ecological thinking involves the interconnectedness of the world which is accomplished in the visual language and musical performance—the entire film becomes music, as Longfellow asserts, created out of "the cacophony" of "living, breathing, and sentient matter".[15] Reeves' sound mix of the reverberations and harmonics of the natural world along with Skúli Sverrisson's exhilarating score is superlative.

Not surprisingly this thinking led to Reeves' next work *Landfill 16* (2011), a short film whose 'canvas' was filled with the extensive outtakes of *When It Was Blue*, which she realized would end up in a landfill. Reeves began by burying the footage to let it decompose with the help of enzymes and microorganisms, after which she hand painted each frame to create an exquisite short abstract film. The film is also "a meditation on the demise of the beautiful 16 mm medium and nature's losing battle to decompose the relics of our abandoned technologies and productions".[16]

Tactility

Both Wieland and Schneemann have inspired an international generation of feminist filmmakers who have been experimenting with expanding the frame through multimedia forms of making. They are committed to working with celluloid films through ecological and materialist scaffolds that recall a history of women's crafts. Jodie Mack (USA), Mary Stark (UK), and Lindsay McIntire (Inuit, Canada) are inventing unique forms of filmmaking, exploring crafts such as stitching, quilting (textiles), and beading. Each of them reflects on a history of these crafts, researching specific kinds of practices, communities, and techniques that each draws upon for their film works. With this in mind, we turn to Kelly Egan's two quilt films which have been dedicated to Wieland and Atkins respectively. The first, *c: won eyed jail* (2005), an anagram of "Joyce Wieland", was directly inspired by Wieland's films, paintings, and media art. For Egan, quilt films offer an alternative language and history to the male-dominated structural film genre as described by P. Adams Sitney.[17] Neither Wieland's nor Schneemann's films were included in the structural film canon precisely because they focused on the personal, on feminine iconography, the messy elements of nature, and the more than human life. Quilts, Egan maintains, operate as a form of feminine visual language: "Quiltmaking with 35 mm is an extension of the frame, of the filmstrip, of the surface". But it is also an editing technique that draws on the tradition and long history of women's collective utilitarian craft. In this way, quilt films conjure "a feminist experimental form of narrative" that is made in and through the materiality of diverse elements that come together to create the work.[18]

Egan's second quilt film is most extraordinary. *Athyrium Filix-Femina* (2016) is a film devoted to Anna Atkins. Egan uses Atkins' 1843 cyanotype recipe to make the film—and faithfully follows the technique, using ferns, vegetation, flowers, and a found 35 mm film which she contact-prints using the cyanotype solution. In order to make the quilt she coats 35 mm clear leader with cyanotype emulsion (Atkins' original recipes with the addition of Knox gelatin), lays the filmstrips into "quilt patches" and directly prints plant materials onto each patch. Once the image is stable, she cuts and edits the pieces together to form a queen-sized quilt (Fig. 6.3).

The end result is that the quilt film exists in several forms. They exist to be projected and also can be dismantled as a film and made into wall hangings for the home or gallery.

The film images all have the cyan blue coloration that characterizes the cyanotypes. The optical soundtrack is tied to the emulsion, the sprockets, and the splices. The combination of the fern fragments is intercut with a found film depicting a young girl being terrorized by a boy who holds a spider in a glass jar up to her face. The scene is repeated alongside the ferns which come to resemble spiders in their beautiful twisting shapes. The juxtaposition is striking throughout *Athyrium*—evoking the natural history illustrations to which Atkins contributed and the male-dominated institutions of science which excluded her. Spiders are generally seen as feminine emblems—weavers of worlds. In the found footage film, the spider is encased and used as threat, a means to frighten the young girl who does not know her own power. The cyanotypes function as an outside space that both overwhelms and liberates the girl who, like the spider, is also encased in the window of the cinema. The flora (cyanotype impressions on celluloid) seem to act as a protector for the young protagonist, opposing the abusive instant in a kind of quilt-covered rescue.

Egan's vast knowledge as a media archaeologist/preservationist is evident in the construction of the soundtrack. The sound was derived from the patterns of the plant materials as they pass over the optical head of the projector. On second viewing/listening, we were able to understand what we were feeling viscerally from the soundtrack, during the first screening. The sharp jabs of sound created by the more angular details of the cyanotype plants work as a metaphor to the aggressive visuals of the imprisoned spider and the boy's attack. The sound and the images gradually become more calming and tactile, as the beautiful cyanotype floral

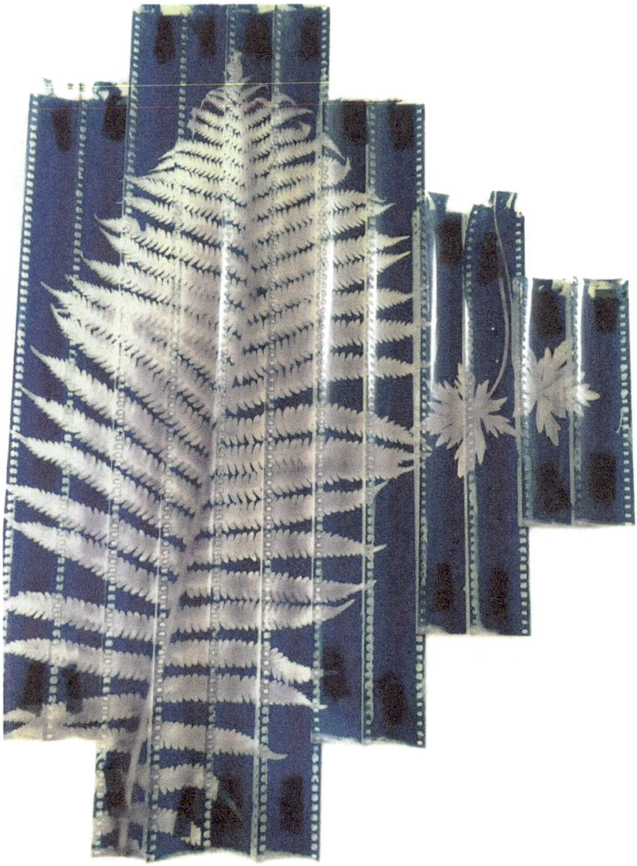

Fig. 6.3 Kelly Egan, cyanotypes for *Athyrium Filix-Femina* (2016) © All rights reserved by the artist/Courtesy of Light Cone

patterns seem to lead the girl to a more comfortable understanding of the open worlds before her. Egan's training as a film preservationist has made her especially committed to working with celluloid and to expanding the platforms and situations which support work using analogue media.

Earth Weavings

In this same vein, the Odeimin Runners Club (Debbie Ebanks, Rebeka Tabobondung, Adrian Kahgee), part of a collective project *Everything I touch, I change*, is committed to using film as a collaborative social practice by experimenting with plant-based processing imbued with the group's personal histories. The Runners ask: "Is it possible to recalibrate contemporary hemispheric relations outside of colonialism such that the settler position is one of being good guests with responsibilities to the hosts? Can we live more gently with the land and all the living beings that depend on it?"[19] The artists propose to exchange works and share knowledge (rituals, ecologies, art) tied to each other's territories—Saugeen First Nation (Kahgee), Wasauksing First Nation (Tabobondung), both located on Turtle Island (specifically Ontario, Canada), and Jamaica (Ebanks) to materialize the "interdependencies of all my relations".[20] In her book *Decolonizing Methodologies*, Linda Tuhiwai Smith has explained that "All my relations" has specific meanings within Indigenous communities:

> The arguments of different indigenous peoples based on spiritual relationships to the universe, to the landscape and to stones, rocks, insects, and other things, seen and unseen, have been difficult arguments for Western systems of knowledge to deal with or accept. These arguments give a partial indication of the different world views and alternative ways of coming to know and of being which still endure within the indigenous world.[21]

Everything I touch, I change, Odeimin Runners' first collective project, has multiple components including films, interactive maps, and augmented reality workshops with Indigenous, Black, and racialized youth from the different territories. Inspired by the novel *Prophesy of the Sower* (1993) by the Afro-futurist writer Octavia E. Butler, the three artists established a "club" that engages with decolonizing methodologies, Indigenous knowledges, and climate ecologies/crises. Making films through an ecological framework that includes their own collaboration, the artists took the first lines of Butler's novel as their starting point: "All that you touch You change. All that you Change Changes you."[22] This is the personal philosophy of Butler's central character Olamina, a young African American woman who conceives of a new religion and a consciousness that understands survival through "autonomous community building". While Olamina looks to rebuild a life on other planets, the Runners propose instead to build communities connected to Indigenous wisdoms. In her artist statement, Rebeka Tabobondung explains that rather than moving out to colonize new plants, the group would re-engage traditional "ceremonies, lodges, offerings, and sacred sites" in order to access the consciousness, knowledges, teachings, and… the messengers that can lend to our survival".[23]

Through workshops with Film Farm (Independent Imaging retreat), the group learned about eco-processing with local plants, as well as the creation of phytograms. This cameraless process recalls Anna Atkins' tactile approach, and proposes a method developed by Karel Doing that lays flowers taken from the land directly onto 16 mm or 35 mm celluloid film. These impressions are deeply beautiful—sometimes taking on the shape and color of the flowers and at other times generating surprising forms and details. This process enables the group to incorporate the multi-layered histories, atmospheres, and temporalities of place into the interpenetrating plant–human materialities of their collaboration. In "Phytograms: Rebuilding Human-Plant affiliations", Doing explains the complexity of these interconnections:

> …plants respond to a variety of different cues, such as moisture, light, gravity, sun direction, barometric pressure, magnetic fields, nutrient gradients, temperature, colour (wavelength), physical contact, fluid flow, electrical current, darkness and the volatile airborne and soluble waterborne chemical signals exchanged within their ecosystem.[24]

The plants and flowers also share nutrients in a communication network that is far more complex than previously understood.

As part of their films, each artist has carefully selected plants, flowers, rocks, and bones tied to their life histories. Adrian Kahgee with *Everything is Right Here* (2021) develops her film with dandelions (a colonial plant) and trilliums (indigenous to the Saugeen territory) to consider their histories (how they came to be on the land) and their medicinal properties. "At the center of my metaphorical bag" writes Kahgee, is "cultural knowledge that is deeply woven into the land."[25] Plants are also connected to understanding the world in terms of change, transformation, survival, and interconnectedness. All three films juxtapose phytograms with historical elements: a jingle dress and dance (Kahgee); the place of a Black settlement connected to the Underground Railroad on Saugeen Ojibway Nation Traditional Territory (*The Traveller*,

Schlums); and a mammoth bone found years earlier by the artist on a beach (*Mammoth Bone*, Tabobondung). All three films bring together inherent knowledge of the land to produce Afro-Indigenous relations by looking to the history and future of the planet. This includes the movements of Indigenous nations across the land and diasporic histories of Canada. The "exploded canvas", coined by Schneemann to describe the integration of her own embodied experience into non-representational actions, comes full circle and expands outward to incorporate deeply integrated and relational temporalities. All three women perform in their films just as they work beyond them to animate forms of community building that connect to their own histories and the land.

This is also why we see so many women working with "eco impulses" create gathering spaces for pedagogies, screenings, and exchanges beyond the films they make. We call these practices 'booming harmonics' because they expand outwards from deeply embedded historical structures, laying the ground for reflexive ecologies.

Notes

1. Scott MacKenzie and Janine Marchessault, *Process Cinema: Handmade Film in the Digital Age*, Montreal, McGill-Queen's University Press, 2019.
2. Larry J. Schaaf, "The Cyanotypes of Pioneering Photographer Anna Atkins", *National Gallery of Canada Magazine* November 26, 2020 [accessed Oct 10, 2021].
3. *Ibid.*
4. Ann B. Shteir, "Gender and 'Modern' Botany in Victorian England", *Osiris*, issue 12, 1997, pp. 29–38, http://www.jstor.org/stable/301897.
5. Annemarie Iker, "Art and Artists", MOMA online https://www.moma.org/artists/231 [accessed October 11, 2021].
6. L. Schaaf, "The Cyanotypes of Pioneering Photographer Anna Atkins", *art. cit.*
7. Johanne Sloan, "Joyce Wieland at the Border: Nationalism, the New Left and the Question of Political Art in Canada", *The Journal of Canadian Art History*, issue 26, 2005, p. 91.
8. Film Farm alumni continue to teach handmade processing techniques. The late Helen Hill's "Ladies Film Bees" program picked up on the common idea of the "Ladies Knitting Bees", offering

an afternoon of shooting and processing in Halifax, Canada and New Orleans, among other locations. Eva Kolcze, Zoë Heyn-Jones, and Terra Jean Long imagined workshop cycles around plant-based processing in Jalisco, Mexico and Gibraltar Point on Toronto Island.

9. Redcat Calarts' Downtown Center for Contemporary Art, "Jennifer Reeves: When It Was Blue", April 26, 2010, online at https://www.redcat.org/event/jennifer-reeves [accessed November 20, 2021].
10. Michael Sicinski, "Spotlight | When It Was Blue (Jennifer Reeves, US)", *Cinema Scope*, August 29, 2009. https://cinema-scope.com/spotlight/spotlight-when-it-was-blue-jennifer-reeves-us/ [accessed October 15, 2021].
11. Dipesh Chakrabaty; "The Climate of History: Four Theses", *Critical Inquiry*, vol. 35, issue 2, 2009, p. 206. For a longer discussion of art and climate change see Janine Marchessault, *Ecstatic Worlds: Media, Ecology and Utopia*, Cambridge, The MIT Press, 2017.
12. Brenda Longfellow, "The Immediate Sensuous: The Process Cinema of Jennifer Reeves" in Scott MacKenzie and Janine Marchessault (eds), *Process Cinema: Handmade Film in the Digital Age*, Montreal, McGill-Queen's University Press, 2019, p. 385.
13. *Ibid.* With "eco impulse" Longfellow is quoting critic Sophie Mayer who suggests that the film's influences come more clearly from Maria Menken or Margaret Tait. Sophie Mayer, "The skin, the aging, the imperfection, the colour, the beauty", *Vertigo Magazine*, Volume 4, Issue 2, Winter-Spring 2009.
14. Carolee Schneemann, *More Than Meat Joy: Performance Works and Selected Writings*, New Platz (NY), Documentext, 1997, p. 167.
15. B. Longfellow, "The Immediate Sensuous: The Process Cinema of Jennifer Reeves", *art. cit.*, p. 256.
16. Jennifer Reeves, "Landfill", *The Film-Makers' Cooperative Catalogue, New American Cinema Group.* https://film-makerscoop.com/catalogue/jennifer-reeves-landfill-16 [accessed October 10, 2021].
17. P. Sitney Adams, *Visionary Film*, 2002, Oxford, Oxford University Press, p. 347.
18. Kelly Egan, "*c: won eyed jail*: Sewing the Divide between Form and Function, Art and Craft, Precept and Concept through the Feminization of Structural Film", *Intersections 2006: Hybrid Identities*,

York & Ryerson Universities, 2005 (Joint Graduate Programme in Communication and Culture), conference presentation.

19. These words are taken from email correspondences with the artists.
20. *Ibid.*
21. Linda Tuhiwai Smith, *Decolonizing Methodologies*, London and Dunedin NZ, Zed Books and University of Otago Press, 1999, p. 84.
22. Octavia, E. Butler, *Parable of the Sower*, London, Headline Book Publishing, 2019.
23. Rebeka Tabobondung, Artist Statement for *Mammoth Bone* (2021). https://cfmdc-tv-preview.netlify.app/works/mammoth-bone [accessed October 15, 2021].
24. Karel Doing, "Phytograms: Rebuilding Human-Plant Affiliations," *Animation: An Interdisciplinary Journal*, vol. 15, issue 1, March 2020, p. 26.
25. Adrian Kahgee, Artist Statement for *Everything is Right Here* (2021). https://cfmdc-tv-preview.netlify.app/works/everything-is-right-here [accessed October 11, 2021].

CHAPTER 7

The Messy Realities of Multi-species Cinema

Karel Doing

Utopian visions of a better world often aim for perfection, an end state of perpetual equilibrium, starring humanity as a brilliant contributor to all that matters. But the alarming actuality of ecological decline calls for something different. As Donna Haraway has pointed out, we have to "stay with the trouble" and argue against seductive narratives recounting "edenic pasts and apocalyptic or salvific futures".[1] Instead she proposes a radical being in the here and now, acknowledging our problematic, contradictory and almost nightmarishly complex present. On a similar note, Timothy Morton speaks about "jumping in the mud" as an antidote to the Hegelian notion of the Beautiful Soul who "sees the world as evil and itself as pure. Or the other way around: evil me, pure world (humans are a viral stain)".[2] Morton's mud is both real mud and a form of conceptual stickiness that needs to be scrutinized and appreciated rather than being washed off.

The notion of the untidy is beautifully addressed by polymath Gregory Bateson in a conversation between him and his young daughter. She

K. Doing (✉)
Oxford, England
e-mail: mail@kareldoing.net

E. Della Noce and L. Murari (eds.), *Expanded Nature*, Experimental Film and Artists' Moving Image,
https://doi.org/10.1007/978-3-031-70729-2_7

starts by asking why things get in a muddle, reflecting on the mess in her own room, either caused by herself or her parents. Bateson comes to the conclusion that: "it's just because there are more ways which you call 'untidy' than there are ways you call 'tidy'".[3] This remark can be extended to the world outside his daughter's room, showing how humans always attempt to organize the world in a particular way that is perceived as being 'tidy'. Once the superiority of tidiness is being questioned, the importance of muddle or mud can be acknowledged, simply as a form of organization that is beyond our grasp. This does not mean a 'laissez-faire' attitude that embraces complete disorder, but rather a fascination with other forms of tidiness that exist beyond our event horizon.

Experimental cinema has long rejected the 'perfect' illusion created by mainstream media by embracing the material qualities of the medium itself in a reflexive way. This concept finds its origin in Brechtian theatre and has been further developed by scholars such as Jean-Louis Baudry[4] and Laura Mulvey[5] in their groundbreaking publications that have gained much acclaim across cultural studies. Similar ideas are applied in experimental cinema by Peter Gidal as a form of critique against illusionism and false notions of representation.[6] But the subsequent arguments of these scholars and practitioners are concerned with the dominance of capitalism and patriarchy rather than anthropocentrism. However, media reflexivity as a concept is still relevant from an ecocritical point of view. Cinema is very much an art of fastidious cleanliness: avoiding dust, scratches, glitches, or any other 'disturbing' interferences is seen as an essential prerequisite to professionalism. Simultaneously, the very basic material elements of both analog and digital cinema (emulsion, silver-halides, electrons, silicon) are undeniably part of the chaotic and unpredictable world that we call 'nature'. For me the question is how to embrace and represent such a disorderly world.

The problem is twofold: it exists on a material level (the film as an object) and within the semiotic sphere (cinematic language). Film culture, even experimental film and video culture, is highly concerned with human stories. When nature is brought into the equation, it functions either as a backdrop, an object of study or as a symbolic reference that tells us something about the inner state of the artist. Animals, plants, meteorological phenomena or the elements seldom have a voice of their own. Similar to the "to-be-looked-at-ness" and resulting objectification of women as formulated by Mulvey, any creature or object beyond the human is denied meaning-making capacity in those forms of cinema that are concerned with reality. Sentient animals, plants and rocks only appear as metaphors, not to be confused with human subjects. The concept of nature as an active participant in the semiotic sphere has received only marginal attention in moving image culture.

An interesting exception can be found in the practice of 'weathering' film. Already exposed and developed film is subjected to the elements, either by subjecting the film for a prolonged amount of time to wind and rain, or by submerging or burying it. After 'harvesting' the deteriorated strips, the remaining images are re-contextualized by the filmmaker. The take-away message is often concerned with the lack of stability of photochemical film as a medium, more symbolic forms of (personal) loss and destruction, or rather the opposite, as a celebration of the prevailing human spirit captured in representational images that are resisting their partial cinematic decomposition. The German experimental filmmaker Jürgen Reble has developed this idea further by foregrounding chemical reaction. In his films and expanded cinema performances images emerge and deteriorate by means of a process that can be experienced *in* action or seen mid-action. In his work photographic emulsion and chemical exchanges are active carriers of meaning beyond a mechanistic interpretation.

An affiliated approach can be found in the work of the British-Canadian experimental filmmaker Chris Welsby. Welsby takes a systemic approach, looking for possible ways to connect natural phenomena to media technology (both analog and digital). For example, in his 16 mm film *Anemometer* (1974) this is done by coupling a device that measures the speed of the wind to the speed control of his film camera. In other works the Earth's rotation, ebb and flow and the direction of the wind are used as active participants in the process of making, coupling mechanical or electronic devices to natural phenomena. In the artist's own words: "mind, technology, and nature are not seen as separate things divided along Cartesian lines, but as interconnected parts of one larger dynamic system".[7] Comparable to the active molecular exchanges in Reble's work, in Welsby's films and installations meteorological events are taking on an active, participatory role, enticing the spectators to rethink their relationship to the natural world.

Explorations of the interdependence of the cultural and the natural domain also appear in the form of direct animation in the work of Len Lye and Stan Brakhage. Lye skilfully explores movement in his hand-painted, scratched, and stencilled films. Although his films are often referred to as abstract there is an indexical relation with the natural world. Instead of being abstract his markings are, rather, ideograms in motion. Through intense observation the artist manages to "levitate with curling smoke, scud with the wind-blown leaf, sashay with the reflection of masts on water, shimmy with the flapping flag, glide with the snake".[8] Brakhage, on the other hand, sources his materials directly from nature, affixing moths' wings, seeds, flowers and leaves to the filmstrip. The final result is aimed at the creation of a new experience of seeing, investigating the internal workings of the visual system. In Brakhage's work the abstract and the concrete also coincide but the resulting motion in his films is inventive, seeking to reanimate the natural world within the context of the cinematic apparatus. As such, both filmmakers have contributed in a distinct way to the development of a cinematic language that explores nature beyond objectification.

Arguably, all four filmmakers are looking at relations beyond the human, investigating how chemical reactions, weather phenomena, animals and plants can participate in the production of moving images. This proposition is not without risk, as it takes us closer to a line of thought that is seen as fundamentally wrong. Professor Arne Johan Vetlesen succinctly delineates the issue that is at stake:

> Being objects, not subjects, one does not enter into contact with plants or animals. And should anyone do so—perceive and treat nonhumans as persons on an equal standing with oneself qua human—it would not be regarded as an innocent mistake, a neutral error, for which one may be easily excused.[9]

Nothing less than conflicting worldviews are competing here for truth, a Cartesian or reductionist truth in which humans are separate from all other living creatures in their ability to interact cognitively with the world, versus a panpsychist truth in which consciousness is an emerging property[10] or an animist truth in which animals, plants, some objects and certain phenomena can be regarded as persons "because they relate and communicate".[11]

While considering the relation between the natural and the cultural domains a compelling fact is that early photography, botany and plant chemistry are related in a number of ways. Pioneering British photographers such as John Herschel, Anna Atkins and Henry Fox Talbot had an interest in the natural sciences and saw photography as a promising method to produce more accurate representations of plant specimens. But the connection also exists in the material sphere. In 1842 Herschel published a paper describing the use of plant juices to make Anthotypes, a method presented to him by Mary Somerville.[12] A year earlier Henry Fox Talbot had patented his Calotype process, using gallic acid as a primitive developing agent. Gallic acid is a compound that can be harvested from the galls of oak trees, a method already in use for centuries to produce ink. In 1831 the French chemist and botanist Henri Braconnot had managed to isolate pyrogallic acid from tree bark, paving the way for an improved photographic developer formula.[13] So-called 'pyro' developers have a long history and are still in use. A further application of similar plant-based phenolic acids in photographic developers has been popularized more recently by Dr Scott Williams' coffee-based formula, widely known as Caffenol.[14]

I came across this formula in 2013 while working on a practice-based research project focused on ecology and expanded cinema, taking into account that in a genuine ecological context not only the content of the produced artwork but also its material constituents and methods of production are important. In order to connect form, content and method I started looking for ways to use organic material as an active mark-making element. I used expired stock that had lost its value within the film industry and would have been thrown away otherwise. My aim was to 'grow images' working directly on the film's emulsion. After analyzing Williams' formula I concluded that it might be possible to use leaves and petals in their entirety rather than extracting their juices. The first experiment with mint leaves soaked in vitamin C and soda immediately yielded an appealing result and I continued to evolve this method. I chose to use a variety of plants from the metropolitan environment in which I was living at the time. My concern was to look at my cultivated surroundings as a different form of wilderness by giving priority to weeds rather than cultivated species. Dandelions turned out to be particularly suitable while working on 35 mm film (Fig. 7.1).

But in order to create animated movement directly on 16 mm film I had to find smaller (wild) plants that would coincide with a single film frame. I started to experiment with ground elder, wood sorrel, herb robert and wild onion, creating repetitive patterns small enough to appear in projection. I reworked the resulting sequences aided by an optical printer while altering exposure, projection speed and colour. This allowed me to investigate the form, rhythm and texture of the chemical traces that

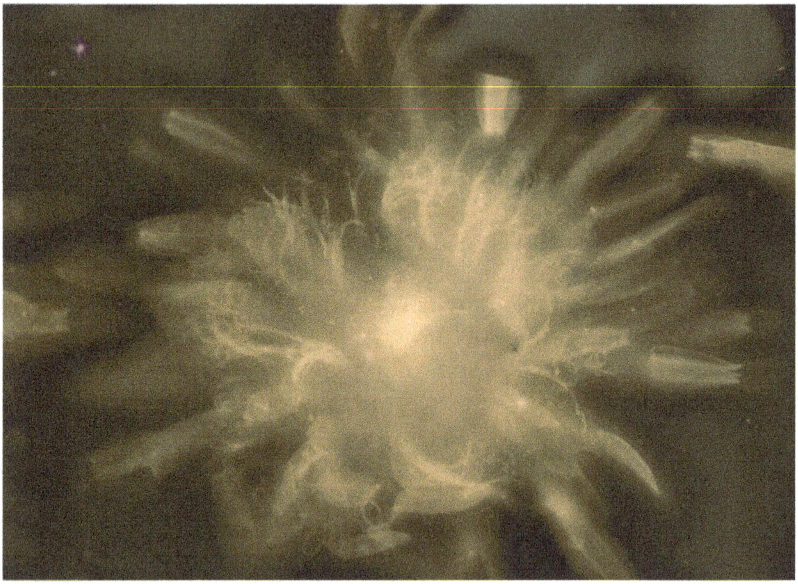

Fig. 7.1 Karel Doing, *Dandelion* (2016) © All rights reserved by the artist/ Courtesy of Light Cone

the plants had left on the film in more depth. I also developed an installation, weaving fishing line through the sprocket holes of the film and suspending the film strips between floor and ceiling in the middle of a gallery. Both projects were concerned with the rich chemical structure of plants as well as their repetitive and ornamental appearance. The aim was to develop a form of notation derived from the plants, resisting the notion of representation. I followed the inscriptions made by the plants rather than forcing the plants to submit to my will.

In July 2019 record temperatures were reached during the second heatwave of the year. I continued working with phytography as a way to record the climatic circumstances. I started composing intricate patterns of plant material on sheets of photographic paper in bright sunlight, a time-consuming process. Subsequently, the plants were wiped off quickly in one decisive gesture and the image fixed. As a result the chemical traces vary in intensity, depending on the amount of time that the plant has remained in contact with the photographic emulsion. The images

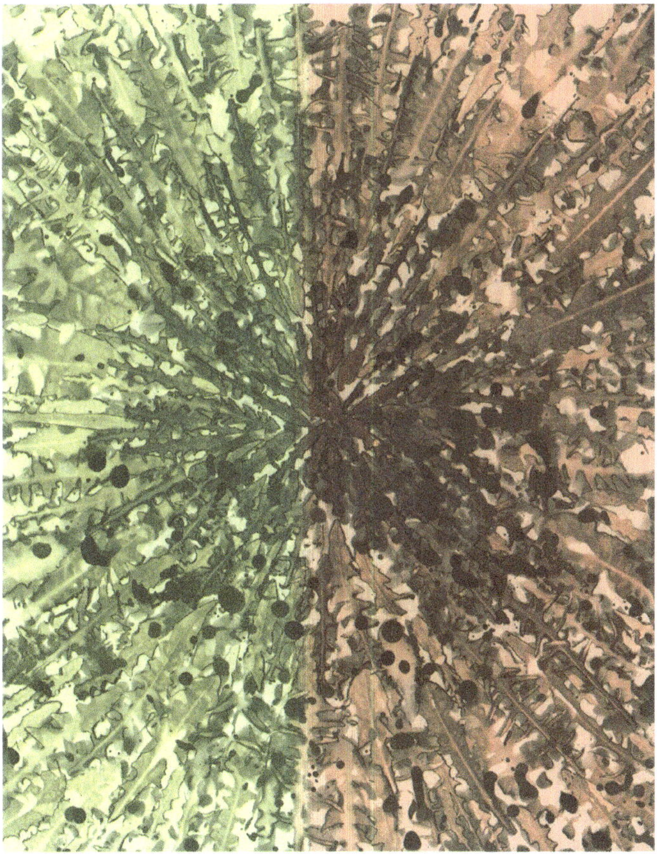

Fig. 7.2 Karel Doing, *Time Lapse 3* (2019) © All rights reserved by the artist/ Courtesy of Light Cone

combine an indexicality in terms of place (the wild plants growing in the yard), in terms of climate (the intensity of sunlight) and in terms of time (the duration of the exposure). The lack of certainty over the outcome of this process was precisely what interested me, aiming to explore the agency of plants and climate in the context of this relatively simple mark-making technique (Fig. 7.2).

By approaching plants as active participants in the process of image making, the idea of a shared space between human and plant was becoming increasingly tangible. Accordingly, it seemed like a logical step to bring other people into this atypical social space as well. Accelerated by the increasing visibility of my work through screenings, exhibitions and talks, I started receiving invitations to give workshops. These workshops took place in a number of contrasting rural and urban settings. Each setting brought me in contact with a particular ecosystem, some partially wild, others highly cultivated. This led to experiments with a much greater variety of plants, weather conditions and film stocks. I gathered the resulting film strips during the following year and brought these together in an eight-minute film simply titled *Phytography* (2020). By approaching the footage as a found object, the creativity and expressive qualities of the plants themselves are highlighted, thus inviting the spectators to discover the subtle differences in tone, texture, form and rhythm.

Written down in this way, the described steps might appear systematic, leading toward a pre-planned destination. But a better analogy would be one of wayfaring, exploring different routes while always remaining both open and alert to an encounter with the unfamiliar. This brings me back to my earlier references to trouble and mud; my practice is much more informed by contingency and curiosity than by a desire to achieve an advantageous outcome. Similar to human relationships, the relationship with plants, other living beings and the earth is never finished and will only remain healthy by means of continuous dialogue and reinterpretation. Plants are not able to speak but they certainly do express themselves through a multitude of methods that in many cases do coincide with human perception. The only thing one needs is a willingness to value this. A meaningful dialogue can start from this point of departure, exploring mutualistic ways of habitation.

Maybe the best articulation I have found to demonstrate this fragile process of exchange is through a series of experiments in which I have combined 16 mm camera footage with phytography. I first filmed a series of portraits of my friends and colleagues. This footage was processed following a so-called reversal process and plants were brought in contact with the latent image during the final step, acting as second developer. Neither the plant nor the human is left wholly intact in this image but both renditions 'consume' each other in an ongoing exchange of elements. Set in motion, such images create a movement somewhere between fight and embrace, analogous to the entanglements of life itself, regardless whether it is human or non-human. This lands me straight back into the mud without the prospect of arriving at a utopian destination. The trail that I have followed does not lead to an end point but rather loops and twists, never repeating the exact same pattern twice.

To go back to the beginning of my investigation, while experimenting with different possibilities to 'grow an image' I gathered sludge from Deptford Creek, one of the few remaining tidal rivers in London that have not been culverted. The creek and creekside are partially designated conservation areas, confirming the ecological significance of the stream's mudbanks. These silt banks appear as a messy anomaly within their concrete, glass and steel surroundings. I impregnated expired 35 mm film stock with this sludge for about 24 hours, eventually washing and cleaning the film material carefully. The emulsion cracked and flaked but also a red-and-brown, cloudy, meandering pattern emerged, resembling pictures of the human brain. I used this footage in my film *Wilderness Series* (2016) alongside sections that were produced with the help of various organic and inorganic mixtures that were brought in contact with the emulsion. I have described this in more detail in my PhD thesis, calling these images 'organigrams'.[15] This process was purely experimental, a jump into a maelstrom of chemical reactions. Yet, this project was key to everything else that I have described and analyzed above. Not only my understanding of, and relative control over, organic/photographic processes, but also my receptivity to plants as lively beings and the ecosystemic properties of soil have grown. The importance of such a relationship between humans, plants and the earth is powerfully expressed by writer and botanist Robin Wall Kimmerer: "The story of our relationship to earth is written more truthfully on the land than on the page. It lasts there. The land remembers what we said and what we did."[16]

Notes

1. Donna Haraway, *Staying with the Trouble: Making Kin in the Chthulucene*, Durham, Duke University Press, 2016, p. 1.
2. Timothy Morton, *Dark Ecology: for a Logic of Future Coexistence*, New York, Columbia University Press, 2018, p. 130.
3. Gregory Bateson & Mary Catherine Bateson, *Steps to an Ecology of Mind: Collected Essays in Anthropology, Psychiatry, Evolution, and Epistemology*, Chicago, Chicago University Press, 2000, p. 5.

4. Jean-Louis Baudry, "Ideological Effects of the Basic Cinematographic Apparatus", *Film Quarterly*, issue 28, Winter 1974–1975, pp. 39–47, https://doi.org/10.2307/1211632.
5. Laura Mulvey, "Visual Pleasure and Narrative Cinema", *Screen*, issue 16, 1975, pp. 6–18, https://doi.org/10.1093/screen/16.3.6.
6. Peter Gidal (ed.), *Structural Film Anthology*, London, British Film Institute, 1978.
7. Chris Welsby, "Films and Installations A System View of Nature" in Jackie Hatfield (ed.), *Experimental Film and Video: an Anthology*, Eastleigh, John Libbey publishers, 2006, p. 30.
8. Luke Smythe, "Len Lye: The Vital Body of Cinema", *October*, vol 144, issue 76, https://doi.org/10.1162/OCTO_a_00141.
9. Arne Johan Vetlesen, *Cosmologies of the Anthropocene: Panpsychism, Animism, and the limits of Posthumanism*, London, Routledge, 2019, p. 157.
10. Philip Goff, *Galileo's Error: Foundations for a new Science of Consciousness*, London, Rider, 2019.
11. Graham Harvey, *Animism: Respecting the Living World*, London, Hurst & Company, 2019, p. 98.
12. John F. W. Herschel, "On the Action of the Rays of the Solar Spectrum on Vegetable Colours, and on Some New Photographic Processes", *Philosophical Transactions of the Royal Society of London*, issue132, December 1842, https://doi.org/10.1098/rstl.1842.0013.
13. James Riddick Partington, *A History of Chemistry, Vol 4*, London, Macmillan, 1964, p. 251.
14. Scott Williams & Technical Photographic Chemistry Class, *A use for that last cup of coffee: film and paper development*, DCCT, September/October 1995, http://www.rit.edu/~andpph/articles.html.
15. Karel Doing, *Ambient Poetics and Critical Posthumanism in Expanded Cinema*, PhD dissertation, University of the Arts London, 2017.
16. Robin Wall Kimmerer, *Braiding Sweetgrass: Indigenous Wisdom, Scientific Knowledge and the Teachings of Plants*, Minneapolis, Milkweed Editions, 2013, p. 341.

PART II

Cooperating With the Living

CHAPTER 8

How to Look at Plants?

Chris Dymond

In Britain, five people were vital for photography's development: Anna Atkins, who made *Photographs of British Algae* (1843–53),[1] the first book to be illustrated entirely with photographs; Anne Dixon, who, with Atkins, co-produced *Cyanotypes of British and Foreign Ferns* (1853)[2]; William John Herschel, who worked to inject colour into the photographic process by experimenting with plant juices; Mary Somerville, who introduced Herschel to the idea; and William Henry Fox Talbot, whose *The Pencil of Nature* (1844)[3] was the first commercial book to include photographic illustrations.

As amateur botanists, all aimed to produce easily duplicable, exact depictions of plants, or to make herbaria: collections of preserved specimens for botanical study. Thus photography arrived burdened with the visual prejudices of previous centuries. Talbot's print, *Bryonia dioica* (1839), featuring a single cutting of English wild vine, is revealing, not least because of its near symmetry to, for example, Leonhart Fuchs's *Anagallis mas*, a woodcut of a scarlet pimpernel in Fuchs's 16th-century

C. Dymond (✉)
Queen Mary University of London, London, England
e-mail: chris.e.dymond@gmail.com

E. Della Noce and L. Murari (eds.), *Expanded Nature*, Experimental Film and Artists' Moving Image,
https://doi.org/10.1007/978-3-031-70729-2_8

herbal, *De Historia Stirpium* (1542).[4] Talbot centralizes and sets the vine upright, its body contorted, its stem, leaves, and spiralling offshoots artfully splayed. Made to complement Talbot's plate, the vine is warped to serve an other's view. Cut from its roots, it is surely dead. Cast against a blank background, the vine's world is inconsequentially erased. Deracinated, flattened, isolated and frozen, Talbot, snubbing plants' dynamic subjectivity, stripped *B. dioica* of agency visually, and methodologically.

But—wait. Aren't plants mere instruments? Ancient wisdom, from Plato, to Aristotle, to the Bible's first sentences, would agree. Plants, says Genesis, are simply "pleasant to the sight" (Gen. 2:9) and "good for food" (*ibid.*): "meat" for animals of every type (Gen. 1:29–1:31); or, as Xavier Bichat, in 1801, states: "the vegetable is only the sketch, or rather the ground-work, of the animal".[5] A rising concert of voices now disagree. Martin Krampen, advancing Jakob von Uexküll's work,[6] devised "phytosemiotics",[7] studying plants' ability to enact semiosis—trade meaningful signs—via their iconic body. Now, plants are seen to share signs through their body and their internal chemistry; or, phytochemistry. And Suzanne Simard calls phytochemistry a vegetal "language".[8] Moreover, Eduardo Kohn, building, too, on Uexküll, argues that if semiosis is present, so is life; that is, subjectivity.[9] Plants—a who, not a what—exceed instrumentalization. Their lives are comparable in depth to, but distinct from, our own.

A red thread, sieved through early photography, ties recent ways of filming plants to ancient botanical interventions. Talbot enacted a unidirectional gaze and established the conventional relationship of camera and plant which remains largely unchanged. Imagery may now move, but plants' treatment is thematically symmetrical. The two foremost ways of filming plants still figure plants from our perspective, and never on their own terms. Each has a landmark case: one amidst the nineteenth century's closure, one amidst Germany's Weimar period.

In Paris, on December 28, 1895, Auguste and Louis Lumière gave their first public screening to a paying audience, showing ten films. The seventh, *Le Repas de Bébé* (1895), is particularly fascinating. In this brief, single-shot and static *actualité*, Marguerite and Auguste Lumière eat with Andrée, nestled between their parents. The trio, centrally framed, invite the gaze. But the crowd, bewitched by the trees rustling behind Marguerite's head and the quivering plant behind Auguste's, grazed over the family's gastronomic adventure, captivated instead by the elusive movement of plants.

In Berlin, on February 25, 1926, Max Reichmann's *Das Blumenwunder* (1926) premiered. *Blumenwunder* opens with some young girls violating a flower bed. Quickly, Flora (Maria Solveg) intervenes, decrying that flowers are, in fact, "alive just like you". Flora's words precipitate a feature-length sequence of time-lapse cinematography where plants flourish and decay in accelerated motion, a procession periodically interrupted by dancers echoing the myriad contortions of *Blumenwunder*'s plants.[10] Reichmann moved Rudolf Arnheim, who, 31 years after its release, yet describes *Blumenwunder* thus:

> *Miracle of Flowers* [*Blumenwunder*] is certainly the most fantastic, thrilling, and beautiful film ever made .... The swaying, rhythmic breathing motions of the leaves, the excited dance of the leaves around the blossom, the almost voluptuous abandon with which the flower opens—the plants all at once come alive and show that they use expressive gestures exactly like those to which we are accustomed in men and animals. ... Plants were suddenly and visibly enrolled in the ranks of living beings. One saw that the same principles applied to everything, the same code of behavior, the same difficulties, the same desires.[11]

The rhythms of plants, diverging from ours, gnaw at our schemas and test our ability to let be. Vexed by plants, Aristotle committed large tracts of *De Anima*, written around 322 BCE, to their analysis. Mobile things, for Aristotle, are undeniably alive. Yet plants, Aristotle notes, are sessile and momentous; landlocked, they wither and grow. Neither fully inert nor wholly vital, plants are divided. Plants, for Aristotle, claim "a partial kind of soul"[12]; or, as Thomas Aquinas in the 13th-century *Summa Theologica*, says: "life in plants is hidden".[13]

Attuning plants' tempo to ours, cinema solves such enigmas. Making plants thrum to an anthropic beat, Reichmann wields cinema technology to register plants' gnomic mobility. But Reichmann bulldozes plant specificity. Time-lapse bends plants until their cadence harmonizes with our pulse, disregarding as inconsequential or non-existent the subjective view of that which it surveys. Subordinating plants, time-lapse entrenches human exceptionalism.

Maximally measured, plants court another trajectory. Apparently, synchronicity with the vegetal is attained through extended duration.[14] Though the shortcomings of going slow have been detailed,[15] a key issue remains inadequately addressed.

Slow media fail to enter the eerie labyrinth of more-than-human subjectivity. Where time-lapse, as David Lavery says, "thicken[s] becoming",[16] foregoing the splice is held to thicken a sense of the real. The world reveals itself amidst editing's ellipses; reality congeals when left uncut. But—which reality? Or—whose? Like its accelerated counterpart, going slow, when used alone, replicates our perspective and revels in the fantasy that no other view exists, that reality fully coincides with human animals' register. Both modes, embodying a definition of plants over two millennia old, enact the anthropomorphic gaze.

The vine crushed beneath Talbot's plate bears cinema beyond the anthropomorphic gaze. Talbot's image is, in fact, no photograph, but a photogram: an image made by laying material on an impregnated surface and lighting it. Odd marks line the vine's stem. These were caused by phytochemistry interacting with Talbot's emulsion. A writerly agency owned by plants was leaking through the gaps of Talbot's control, a final rebellion enunciated by the body of the oppressed. Such carvings offer a departure from standard ways of looking.

Following plants, contemporary filmmaker Karel Doing swerves off-piste, advancing phytography. Many plants contain polyphenols, molecular packets similar to those in photographic developers. Appropriately handled and placed on a chemically infused surface, plants can act as a developing agent. Phytograms convey an intricate view of a plant's body and phytochemical reactivity. Phytography, says Doing, "translates a plant's experience of the world into an image that is legible for humans".[17] To see a plant's body and phytochemical reactivity make an image is to watch a plant enact semiosis and make art. A portal to plant–human animal "communication", phytograms are "a bridge between … two forms of subjectivity".[18]

Doing's In Vivo (2021) inhabits, as Doing says, a "messy reality".[19] A dense tapestry, *Vivo* comprises various segments which coalesce as a messy construct. Disparate registers everywhere interpenetrate. Toying with language, Doing folds human animals into a wider constellation. Early on, a human animal—Uta Paszkowski, a plant scientist specialized in mycorrhizal networks—addresses the camera directly. Uta's gestures are warm but Uta's vocalizations are incomprehensible. The original soundtrack has been removed, replaced by Anna Homler "speaking in tongues",[20] their indecipherable vocalizations contradicting the image.

Vocal but silent, human animals, stripped of verbal language, the tool used to underpin their exceptionalism, are removed from a high perch.

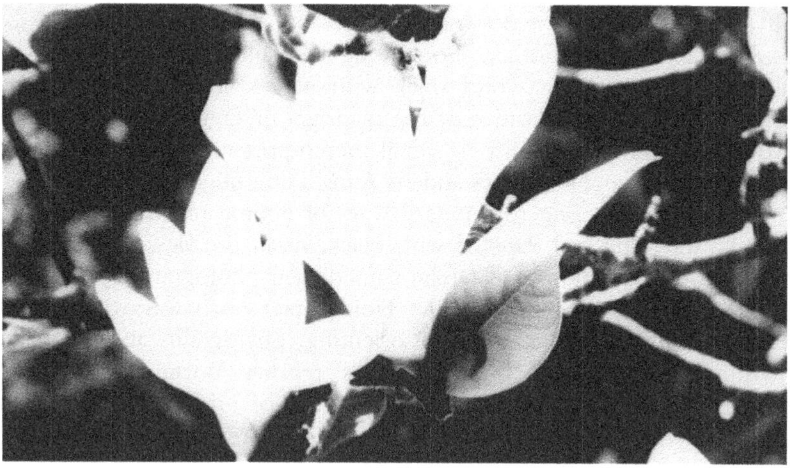

Fig. 8.1 Karel Doing, *In Vivo* (2021) © All rights reserved by the artist/ Courtesy of Light Cone

But plants are seen to signify, and raised. After Uta's performance, an extended phytography sequence occurs. Vegetal figures dance against a metamorphic background. Then, images displaying some leaves and buds appear, albeit with their contrast heightened and spectrum distorted. Scanning as a blinding white, they signal a presence which, though visible, remains partially impenetrable. Phytography, reifying plant subjectivity, renders plants available to sight but unassimilable to our viewpoint. Phytography highlights vegetal recalcitrance (Fig. 8.1).

Later, phytograms showing leaves streaming sideways appear. Immediately, the frame rate falls below the threshold to produce fluid motion, diverging from our rhythm. Then, some leaves' vascular networks emerge. Now, the proximity of viewer to filmed object is tight. Each frame includes only portions of a leaf, derailing a sense of scale, blocking human animals' tendency to comprehend and map. A cut, and more phytograms arrive. A plant's head darts around the screen, evading inertia. Next, some leaves bisect the frame in various directions, their bodies incoherently overlapping, phasing in and out of legibility.

During these segments, Doing's soundtrack, scored by Gareth Davis, oscillates between a nightmarish volatility and a gentle playfulness, coalescing as a musicality both repellent and inviting, signifying that phytography scans as a destructive force in one register and a beautiful enterprise in another. For though derailing anthropocentrism, phytograms envision a world thick with life. Moreover, employing a polyphonic rhythm indicative of the madrigal or fugue, Davis musically evokes phytography's ability to signal the co-presence of discordant but overlapping temporal regimes and, as Anna Tsing might say, lifeways.[21]

Four shots adhering to standard ways of filming plants follow. Such images, soft blue and dappled with black, recall the views of a former century. Indeed, Atkins and Dixon named their photograms cyanotypes, typifying their images' azure hue. Doing stages two ways of imaging plants side by side. Jittery and disorienting, phytograms abide by odd rhythms, deviating from human animals' register. Warmly inviting and curiously alien, phytograms slide in and out of our world. Following plants, phytograms diverge from ways of gazing reaching back, at least, to the sixth century.[22]

One sequence summarizes *Vivo*. Early on, images of human animals overlaid with phytograms appear. Spectral plants cascade across anthropic faces. Phytography was employed at a late stage, as a second developer. Doing flips the schema: a phytomorphic gaze surveys human animals. Yet neither view is privileged. Both are messily balanced. Such imagery, methodologically and visually entangled, traces the untidy, intrapenetrating union of human animal and plant, reframing the world as a messy, nested phenomenon.

Filmed in the first person, Miles McKane's *Track Moss By-Pass* (1987) is a more-than-human 'phantom ride'.[23] *Track*'s opening sequence disorients: a dive, from above some canopy, into a river; a feverish roll amidst icy whorls; an ejection onto a dark bank; and, finally, a staggering crawl across a labyrinthine woodland. McKane devises an arthropod's view of the world. A bulging retina, McKane's convexed lens bends the frame. Scurrying along at an inhuman velocity and suffused by a perpetual semi-darkness, McKane blots out standard lodestars. Speculating on the possibility of a more-than-human viewpoint, McKane prefigures Doing's later pursuit.

McKane advances fresh views of plants. Traditionally, the production of plant imagery necessitated actual deracination. Today, deracination is primarily enacted symbolically. Framed as devoid of world, plants are visually deracinated. Consequently, artists entrench the idea that plants have no world; or, if they do, that their having is secondary to our satisfaction. McKane clings to the ground, refusing to cut plants from their roots.

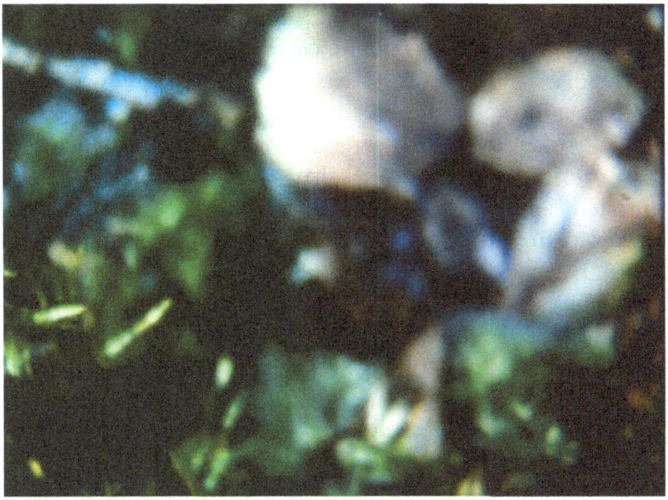

Fig. 8.2 Miles McKane, *Track Moss By-Pass* (1987) © All rights reserved by the artist/Courtesy of Light Cone

And plants' territory registers as a thick, meandering maze. The camera is confused, spinning in place or darting around without clear direction. Individual plants are only partially decipherable, standing out but fading into their milieu (Fig. 8.2).

Born in New Zealand, McKane views plants through the lens of Māori tradition, which, as McKane writes, sees "[p]eople and forests a[s] vitally connected—the same god created both".[24] McKane continues: "This deference towards plants and the common roots of both plant and human is an evidence and is fundamental for me". And shared roots denotes the possibility of communication. Beautifully, McKane states, access to plants is furnished through "hum[ility] in front of the subject". Such quotations capture *Track*: a creative abandonment of a human animal's reality in search of a plant's. *Track* scans as an obscure borderland between two worlds.

A plant's world, McKane shows, is singular but co-built, and an encounter with plant subjectivity enables entry into a messy labyrinth. Constantly blurred, McKane's lens fails to wholly capture. Despite close scrutiny, plants evade analysis. Such blurring, contends McKane, signals a "change of state", a blending of boundaries. Diving downwards, McKane works to disorient and reorient, to "push the viewer ... to question their ... position in the world".[25] Wielding cinema and adopting the arthropod as an intermediary, McKane contracts and expands, pulling focus whilst shunting cinema into a maze devoid of anthropocentric markers.

"The beauty of the world is the mouth of a labyrinth",[26] writes Simone Weil. At the heart of Weil's labyrinth sits God. Finding God, one gets eaten, digested, and spat out. Consumption by God precipitates transformation. Encountering worldly beauty, says Weil, can consume and change. But beauty, for Weil, is, as Anat Pick writes, "neither necessarily pretty nor visually pleasing".[27] Rather, beauty is conterminous with recalcitrance. The beautiful is that which exceeds.

Those who enter Weil's labyrinth are those who acquiesce to an other's sovereignty. A fascinating sequence sits at the end of Charlotte Clermont's video, *How Flowers Never Became a Food Group* (2017). The 80-second uninterrupted shot is vividly metamorphic, the destabilized image testament to Clermont's skill with a video synthesizer. Bands of static bisect an image which, infused with colour, scans as vibrant and sessile, echoing a plant. Within the image sits a purple flower. Quickly, a bee alights on its petals, and, after a busy cavort, disappears, venturing inside the plant only to later slip out—a little greener, perhaps, than before.

Plants call to bees via phytosemiosis. Their voluptuous bodies and alluring scents signal culinary delight or sexual excitement; the emittance of electric fields signifies nutritional non-abundance; a receptivity to vibrations made by bees' wings precipitates the intensification of pollen sugar density. Bees follow phytosemiosis. Landing on a plant's flower, a bee, then, also alights into a labyrinth.

The revelation of phytosemiosis and acquiescence to plant subjectivity manifests, in Clermont's view, as a transformative activity. Clermont's work pinballs with Doing's, for *Vivo*, made with plants, is a maw onto a maze overrun by vegetation. And what else is *Track* but a rip through a dark, leafy labyrinth?

Following such artists, I switch Weil's hungry God for a plant. As that which precipitates metamorphosis, the eater is important. The eaten, after regurgitation, is remade, if only partially, in their eater's image. Through digestion, human animals, as Weil says, can "become different".[28] And plants can digest human animals via cinema. What does plants' eating of the anthropos spark? In living, plants teach many lessons. Rhizomatic, plants intensify their flourishing largely through alliance, not domination; and plants consume, but never without remainder. Plants, rather, creatively share, their bodies the site of a radical hospitality. Even their excretions are gifts.

Following plants coalesces as a gateway onto a more sustainable form of life; or, as Karen Houle writes, "a posthumanist ethicality".[29] But "becoming-plant"[30] requires affective revelation of plant ontology. Cinema, through new views or phytography, offers this. The screen is also a door to a labyrinth at the heart of which resides a plant.

Cinema, too, can become-plant. Cinema's becoming-plant is an offshoot propagated by the adoption of a phytocentric methodology: filming like a plant. This typifies cinema's journey when artists choose to follow plants and make in a way congruent with vegetality. Filming like a plant describes a practice dedicated to letting plants be precisely through the act of capture. Filming like a plant coalesces as a refusal to make art by devouring the world.[31]

At a minimum, the West has devoured plants for over two millennia. To devour is to eat something completely—both ontologically and actually. To devour a plant is to actually imbibe them and to construe them as an instrument, to trace their highest value as their ability to satisfy an other's appetite. Visual media, from ancient herbaria to contemporary cinema, entrenches such views.

Such views get challenged when cinema expands. Cinema's expansion names the medium's movement into territories lacking anthropocentric coordinates. Cinema's expansion dictates a forward momentum which is actually, as Pick might say, a "kind of retreat", a "reticence before [an] other when confronted by [an] other's sovereign existence",[32] A welcoming exertion, it makes room. When control is relinquished and following enacted, the frame shifts from a mirror to a portal beyond our horizon. Cinema's expansion reinvigorates the significance of cinema's designation as a medium.

Though taken for granted, the act of looking at plants is charged and decisive. Plants have been ontologized through art. Dominant styles of gazing have conferred plants to the wayside of our lives. Thinking the question of how to look at plants is one response to such a problem. The other is asking whether it is acceptable to look at all. Their interplay rebounds with the question animating this edited volume. Amidst seemingly unstoppable climate crisis, should we abandon, or expand, cinema?

I advance not looking away, but looking differently. Looking away erects taboos and halts forward momentum. Whereas if looking can do harm, then looking can also serve that which bears the look. But, looking at plants, I want to evade the equation of planetary crisis. If we save plants in order to save ourselves, we block a restorative trajectory. The only valid route to caring for plants is through accepting that plants are.

But what are plants? This question is both impossible to answer and supplemental to a rudimentary counterpart. Namely, that plants enjoy subjectivity. Such knowledge enables learning to let plants be. Letting be is an activity which curtails and recalibrates the self in respect of an other, a simultaneous no and yes. The revelation of plant subjectivity is a question of looking. The question of how to look at plants is partly answered by expanding cinema.

Notes

1. Atkins self-published her book *Photographs of British Algae* and gave the entire work, which comprised thirteen separate booklets, to William John Herschel. This is now held within the Spencer Collection at the New York Public Library. It has been digitized and can be accessed in its entirety via the library's website.
2. Dixon and Atkins self-published their work. Large sections of it have been digitized and are available through the Getty Museum's website.
3. William Henry Fox Talbot, *The Pencil of Nature*, London, Longman, Brown, Green and Longmans, 1844. Also available via the New York Metropolitan Museum of Art's website.
4. Leonhart Fuchs, *De Historia Stirpium Comentarii Insignes*, Basel, Officina Insingriniana, 1542. Fuchs's herbal has been digitized in its entirety by the Hunt Institute for Botanical Documentation. It can be accessed freely via their website.
5. Xavier Bichat, *Physiological Researches on Life and Death*, trans. F. Gold and George Hayward, Boston, Richardson and Lord, 1827, p. 12.
6. For a starting point into Uexküll's exciting work: Jakob von Uexküll, *A Foray into the Worlds of Animals and Humans, with a Theory of Meaning* [1934], trans. Joseph D. O'Neil, Minneapolis: University of Minnesota Press, 2010.
7. Martin Krampen, "Phytosemiotics", in Donald Favareau (ed.), *Essential Readings in Biosemiotics: Anthology and Commentary*, New York, Springer, 2010, p. 266.

8. Suzanne Simard, "Mycorrhizal Networks Facilitate Tree Communication, Learning, and Memory", in Frantisek Baluska, Monica Gagliano and Guenther Witzany (eds), *Memory and Learning in Plants*, New York, Springer, 2018, p. 201.
9. Eduardo Kohn, *How Forests Think: Toward an Anthropology Beyond the Human*, Los Angeles, University of California Press, 2013, p. 16.
10. I thank Matthew Volgraff for this reference. For a fascinating discussion of plants, cinema, and Weimar Germany, see: Matthew Vollgraff, "Vegetal Gestures: Cinema and the Knowledge of Life in Weimar Germany", *Grey Room*, Vol. 72, Summer 2018, pp. 68–93.
11. Rudolf Arnheim, *Film as Art*, Los Angeles: University of California Press, 1957, p. 115.
12. Aristotle, *De Anima*, trans. Mark Shiffman, Newburyport, Focus Publishing, 2011, p. 51.
13. Thomas Aquinas, *Summa Theologica*, trans. Fathers of the English Dominican Province, New York, Benziger Brothers, 1947, b. I, q. 69, art. 2.
14. Scott MacDonald has built and developed this idea through many works. MacDonald's conclusions are now widespread and continue to flourish within the intertwined subdisciplines of ecocinema and slow cinema. Indeed, MacDonald has been pivotal in setting up both subdisciplines. Though our conclusions starkly differ, we both pursue the same inquiry, namely, how to attune cinema to a more than human's reality, and what ramifications such synchronisations set in motion. As a starting point, see: Scott MacDonald, "Towards an Ecocinema", *Interdisciplinary Studies in Literature and Environment*, no. 11, 2004, pp. 107–132.
15. Ryan Conrath, "The Ecological Cut," *Millennium Film Journal*, no. 69, Spring 2019, pp. 84–95; Chris Dymond, "Plant-Filming: Re-viewing Vegetality", *Millennium Film Journal*, no. 73, Spring 2021, pp. 28–38.
16. David Lavery, "'No More Unexplored Countries': The Early Promise and Disappointing Career of Time-Lapse Photography", *Film Studies*, no. 9, Winter 2006, p. 1.
17. Karel Doing, "Phytography", *animation: an interdisciplinary journal*, vol. 15, no. 1, 2020, p. 31.
18. Karel Doing, interview with author, 26 October 2020.
19. Karel Doing, interview with author, 18 April 2021.

20. *Ibid.*
21. Anna Tsing, *The Mushroom at the End of the World: On the Possibility of Life in Capitalist Ruins*, Princeton, Princeton University Press, 2015.
22. Though individuals (Aristotle, Crateva di Mitridate, Pliny, and others) were studying, and apparently drawing, plants as early as the fourth century BCE, it is Dioscoride's 6th-century CE *Vienna Dioscorides*, an illuminated herbal that documented roughly 500 plants, which stands as one of the West's earliest documents to include various representations of plants. Of course, plants pop up in art before Dioscorides: trees appear in rock art, for example, and the Mayan peoples loved to represent fungi—which, though now classified as different to plants, were, during the so-called Classic Maya times, thought of as part of the vegetal world. To learn more about plants in early art, see: Giovanni Aloi, "Introduction: Why Look at Plants?", in Giovanni Aloi (ed.), *Why Look at Plants? The Botanical Emergence in Contemporary Art*, Leiden, Brill, 2018, pp. 1–38.
23. The phantom ride was a late 19th-century film genre akin to the panorama. These films were generally shot by a camera mounted on the front of a locomotive.
24. Miles McKane, interview with author, 1 June 2021.
25. *Ibid.*
26. Simone Weil, *Waiting for God* [1950], trans. Emma Crawford, New York, G. P. Putnam's Sons, 1951, p. 103.
27. Anat Pick, "Vegan Cinema", in Emelia Quinn and Benjamin Westwood (eds), *Thinking Veganism in Literature and Culture: Towards a Vegan Theory*, Oxford, Palgrave, 2018, p. 127.
28. Simone Weil, *Waiting for God*, *op. cit.*, p. 103.
29. Karen Houle, "Facing only outwards? Plant bodily morphogenesis and ethical conceptual genesis", in Angela Kallhoff, Marcello Di Paola and Maria Schörgenhumer, *Plant Ethics: Concepts and Applications*, London, Routledge, 2018, p. 72.
30. Karen Houle, "Animal, Vegetable, Mineral: Ethics as Extension or Becoming? The Case of Becoming-Plant", *Journal for Critical Animal Studies*, vol. 9, issue 1/2, 2011, p. 111.

31. I explore filming like a plant in more depth here: Chris Dymond, "New Growth: To Film Like a Plant", *Ecocene*, vol. 2, no. 1, Summer 2021, pp. 32–50.
32. Anat Pick, "Vegan Cinema", *op. cit.*, p. 128.

CHAPTER 9

Lost Paradises: Animal Spectres and Experimental Cinema

Alice Leroy

Lives of Ghosts (Disappearances, Appearances)

First, there is only darkness. But soon, luminous shards pierce the obscurity, like an organic life rising to the surface of the images. The old emulsion has degraded the shapes and colours, leaving only yellow-and-green-veined traces whose vegetal or purely chemical nature is no longer clear. Suddenly a herd of bison gallops across the flickering strip, as if sprung from a celluloid dream. The dark mass of beasts slowly undulates across the film, fleeing some unknown fate of extinction or returning to haunt the moor from which they were driven. This journey through the matter-time of 16 mm film is imagined by Canadian filmmaker Christina Battle as a passage, a slippage from one world to another, that of the ghost-America of origins, nature doomed to destruction by the very

A. Leroy (✉)
University Gustave-Eiffel, Champs-Sur-Marne, France
e-mail: alice.leroy@univ-eiffel.fr

Institut Universitaire de France, Paris, France

E. Della Noce and L. Murari (eds.), *Expanded Nature*, Experimental Film and Artists' Moving Image,
https://doi.org/10.1007/978-3-031-70729-2_9

people who discovered it as a new Eden—and from one medium to another, through a process of emulsion transfer known as *lift*—hence her film's title, *Buffalo Lifts* (2004). This image and its manufacturing process thus narrate a double migration: that of a species hunted and exterminated as the conquest of the West progressed, erased from North American landscapes to inhabit only our filmic imaginaries, and that of chemical emulsions which alone preserve the memory of vanished worlds on fragile and unstable media. Incidentally, this journey through time also retraces the history of cinema, from the dawn of the 21stcentury, which saw the advent of digital media to the detriment of photochemical film, back to the beginnings of the moving image—with Thomas Edison celebrating in 1894 the most famous of bison killers, William Frederick Cody, known as Buffalo Bill,[1] and seven years earlier, like an epitaph to its disappearance from the plains of the West, the solitary bison of plates nos 699 and 700 of Eadweard Muybridge's photographic atlas *Animal Locomotion* (1887) (Fig. 9.1).

The technological history of cinema overlaps with that of the destruction of natural worlds. This is how, following John Berger, we could qualify the rift that opened up in the nineteenth century and has since continued to widen, between humans and other living beings. In his famous 1977 essay, "Why Look at Animals?", the British writer and art critic established the genealogy and effects of this break: "The 19th century, in Western Europe and North America, saw the beginning of a process, today being completed by 20th century corporate capitalism, by which every tradition which has previously mediated between man and nature was broken".[2] While the theoretical beginnings of this rupture were formulated almost two centuries earlier in Cartesian philosophy and its mechanization of the living, its full realization only played out during the Industrial Revolution, when steam engines, mechanical looms and combustion engines began to replace beasts in the fields and in the cities. Urban expansion then transformed the surrounding countryside, while there began large-scale commercial exploitation of species prized for their skins, furs, wood, horns or oil. In this world entirely subjected to the industrial order, there is no longer room for wildlife, except in designated spaces, reserves and parks, small theatres of nature preserved from the great extermination enterprise.

But the disappearance of animals, initiated in the nineteenth century and uninterrupted ever since, has also seen them appear in other spaces: zoological gardens, natural history museums, domestic habitats or cinema

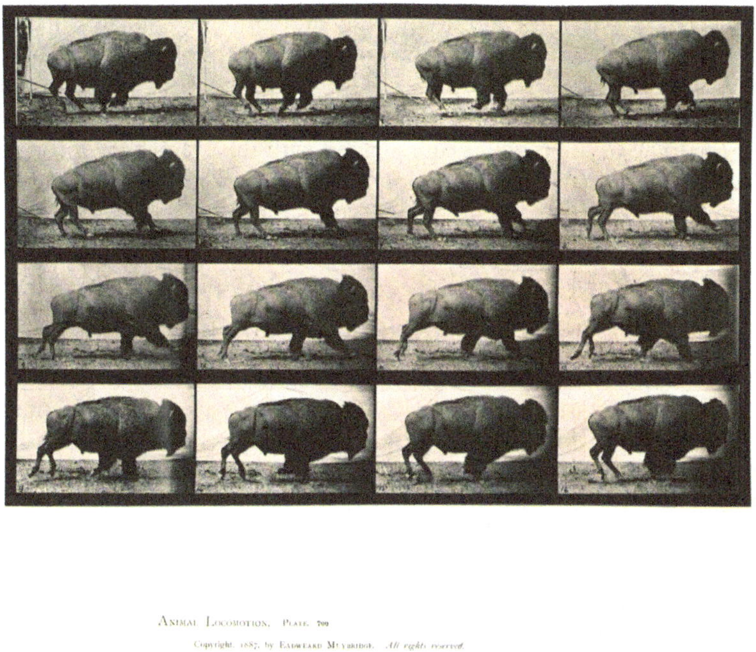

Fig. 9.1 Eadweard Muybridge, *Animal Locomotion. Buffalo; galloping* (1887) © All rights reserved by the artist/Courtesy of Light Cone

screens. Caged, pinned, domesticated and fictionalized, these other beings have become strangely familiar. As animals deserted their natural environments, we indeed began to watch them exist, or pretend to exist, in a solitude that perhaps reminded us of that of our species. "Public zoos", writes Berger, "came into existence at the beginning of the period which was to see the disappearance of animals from daily life".[3] Their genealogy, like that of museums, connects these places to a colonial enterprise of extermination, embodied by the figure of the white hunter bringing back exotic specimens, dead or alive, from his distant expeditions. Unlike royal menageries, intended for the sole benefit of the monarch and his court, zoos are monuments to the glory of empires visited by large and popular crowds, which reproduce through a scenography of nature the full geographical and political extent of the power of the metropolis. The scientific argument that presides over the creation of such gardens fears no aberration, not even the antinomy of a study of these animals outside of any "natural" environment. Even when László Moholy-Nagy, in a 1936 film (*The New Architecture and the London Zoo*), praises arrangements that substitute cages and aviaries with semi-open and transparent habitats, the logic of the gaze remains aligned with the eye of the visitor, who thus enjoys the spectacle of a fiction of nature conformed to the modern imaginary.[4]

Experimental Metaphysics (Revelation, Transmutation)

Let's follow Berger's intuition: cinema establishes itself, with the zoo and the natural history museum, as one of the spaces of representation that makes it possible to conjure up the disappearance of animals at the same time as sanctioning it. More precisely, whereas cinematic fictions re-establish at the mythological level a symbolic continuity between human and other-than-human beings—by endowing beasts with speech for example, as legends and popular tales once did—, a more experimental cinematic tradition explores the material nature of film as a sensory experience of time through the enigma of animal presence. One of Malcolm Le Grice's most famous films, *Berlin Horse* (1970), consists of two loops: the first, which extends over two-thirds of the film, presents a horse trotting around a carousel, and this rotation looped back on itself by the repetition of the same shot, accelerated, slowed down or even reversed, re-enacts the mechanics of the film in the projector. Initially monochrome, the image is gradually tinted with coloured filters until it tips over into a negative inversion, to the rhythm of the repetitive, hypnotic soundtrack composed by Brian Eno. The second, much shorter loop concludes the film on a dramatic note: through the open door of a burning barn, a man ushers out panicked horses (Fig. 9.2).

Le Grice borrowed these shots from a short film by Thomas Edison, *The Burning Stable* (1896), but due to the refilming of the film strip with the optical printer, it is difficult to distinguish them from the ones he shot. The dramaturgy of Edison's film matters little to the filmmaker, it is rather the "affirmation of the projection event as the primary reality"[5] which is important. For these kaleidoscopic loops, with their effects of speed and saturation, effectively produce a deviation of the narrative towards a pure experience of filmic time.

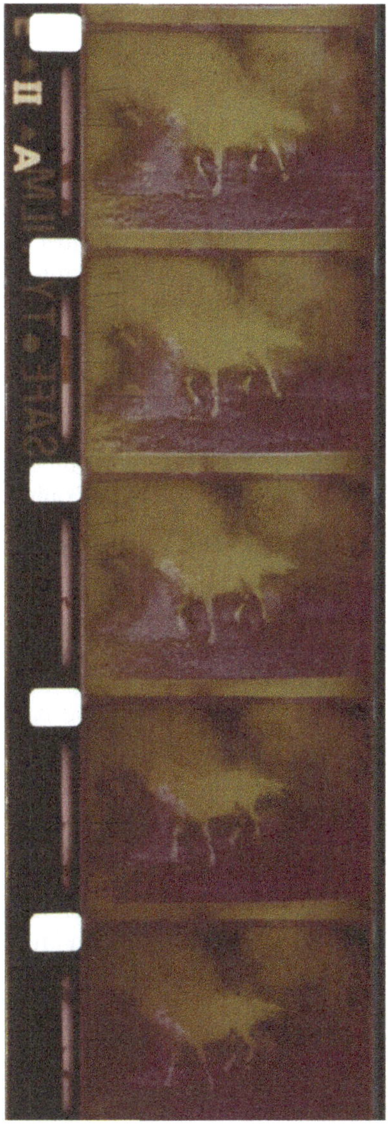

Fig. 9.2 Malcolm Le Grice, *Berlin Horse* (1970) © All rights reserved by the artist/ Courtesy of Light Cone

Others before him, such as Andy Warhol, Peter Gidal, Larry Gottheim or even Michael Snow have explored this field of spatio-temporal equivalences, the film becoming a concrete experience of the space–time of projection.[6] Here, however, this equivalence between filmic and real time is bound to an animal body whose frantic race invokes both the carousel of Muybridge's zoopraxiscope and the first films of the Lumière brothers. *Berlin Horse* is both Muybridgean in its decomposition of animal movement and film mechanics, and Lumièreist in its rediscovery of cinema's inaugural shot through the panicked rush of animals out of the burning stable.[7] The presence of the horses in these images could be considered anecdotal or contingent, but I suspect that this is not the case. In an earlier film from 1967, *Little Dog for Roger*, similarly devoted to the search for spatio-temporal equivalence between the film and its projection, the intermittent silhouette of a small dog in a country landscape already embodied the materiality of the film. But in *Berlin Horse*, the animal not only reveals the film materiality, it constantly transforms it. In the end, it is unclear whether the flames threaten to devour the horses or the film itself. The looped montage of *Berlin Horse* records the inexorable work of time on matter, whether of body or film. The same operation of equivalence between animal movement and the work of time is highlighted in a short film by the Dutch artist Esther Urlus, *Konrad & Kurfurst* (2014). Using a homemade emulsion tinted according to processes gleaned from old photography and cinema manuals, Urlus resurrects the forgotten memory of Konrad Freiherr von Wangenheim, winner, despite a dangerous fall, of a horse race at the 1936 Olympic Games. To do this, she not only reworked images shot by Leni Riefenstahl in Berlin, but also literally filmed horses swimming in water resembling an emulsion bath. Irregular bubbles disturb the bluish surface of the shot, the unstable emulsion and the ghostly silhouettes of the rider and his mount both proceeding from the same spectral alchemy. If *Konrad & Kurfurst* metaphorically embody the tragic heroism of those erased from history, Urlus' homemade emulsions give them another form of existence, detached from any historical context or Olympic legend. They wander indefinitely in the murky thickness of time and the formless matter of film.

Extending Berger's thesis, film theorist Akira Mizuta Lippit suggests that the nineteenth century ushered in less a historical process of mass extinction or replacement of animals by machines, more an ontological change in their being. "[A]nimals", he writes, "never *entirely* vanish.

Rather, they exist in a state of *perpetual vanishing*".[8] In the age of modernity, animals have entered a new economy, not of labour and sacrifice but of image and spectrality. Cinema, more than any other modern technology, has ceaselessly chronicled this "state of perpetual vanishing" by becoming the mechanical unconscious of this loss. "In this manner", writes Lippit, "technology, and ultimately the cinema came to determine a vast mausoleum for animal being."[9] If films form a melancholic imprint of what we have lost—the sacred, mythical or magical bonds that once bound us to other living beings—, they accomplish this mourning operation through a recurring motif: animal death. It is not so much the act of killing as the process of dying that is important here. From Yervant Gianikian and Angela Ricci Lucchi to Peter Kubelka and Cécile Fontaine, many filmmakers have seen in the obscenity of colonial safaris a mirror of the literal or symbolic violence inflicted on the colonized body.[10] But animal death not only acts as a metaphor for other unfilmable violences, it also works at a more subterranean level to viscerally anchor us to this mortal community to which we belong, along with all other organic beings.

Animal Singularities (Specimen, Subject)

In contrast to Le Grice's structural cinema, Stan Brakhage's visionary eye has also focused on animals, particularly to record their disappearance. Before *Mothlight* (1963), a tragic microcosm where life and death are played out on the scale of a beam of light passing through the wings of night moths glued among petals and leaves on the film strip, Brakhage filmed his dog in *Sirius Remembered* (1959). Or rather, he filmed his dog's corpse. His gesture was then less an act of morbid curiosity than an animist rite, probing, with the help of a vibratile and animal camera, "the decay of a dead thing" as he himself recounts.[11] Upon the death of their dog, named Sirius in homage to the main star of the Canis Major (Great Dog) constellation—the Dog Star, which subsequently lent its name to the vast cosmopoetic fresco undertaken by the filmmaker between 1960 and 1965—, Jane and Stan Brakhage refused to bury it. In New Jersey where they then resided, winter had frozen the ground anyway. They instead deposited his corpse in the woods on the outskirts of the city. For months, Brakhage made pilgrimages to the site, where, first as curious observer, but quickly as monomaniac, he filmed the body as it was gnawed away by the cold and insects over the course of the seasons. Unlike many of his compatriots, the filmmaker had never filmed or photographed his dog while he was alive.[12] Despite what its title suggests, *Sirius Remembered* seeks less to preserve the living memory of a pet than to confront the work of time and the progressive erosion of a corpse absorbed by other natural elements. This filmic ritual does not simply involve the work of mourning—accepting at the same time the loss of the body and "the decay of memories of a loved being that had died"—but an animist ritual of magical reanimation[13]: the ecstatic dance of the camera flailing in the surrounding bushes, its upward movements towards the treetops, invoke an animal procession around the corpse whose soul will finally rise to the heavens. "The form was being cast out by … the same physical need that makes dogs dance and howl in rhythm around a corpse"[14] recounts Brakhage. The editing, alternating shots filmed in winter and spring, inscribes the process of bodily degradation in a reversible logic, or at least makes it accessible to our own experience, even though, as the filmmaker reminds us by quoting Wittgenstein's *Tractatus*, "death has no meaning because it is never experienced".[15]

Unlike Le Grice in *Little Dog for Roger*, in *Sirius Remembered*, Brakhage does not film *a* dog but *his* dog. The possessive here refers less to the relationship between master and animal than to the singularity of a relationship established by the camera. A long tradition has linked the images of cinema to the mechanization of the gaze and the objectification of the bodies. This has had the effect of obscuring the fact that cinematic images can also transform any object into a subject. When, in the mid-nineteenth century, a new epistemological paradigm assigned a "mechanical objectivity"[16] to recording devices, with the aim of removing all human agency from representation, it led to the production of a new type of serial images. These mechanical images focused their attention on specimens rather than ideal types, and in their utopian quest for infallible objectivity, they also gave rise to aberrations and singularities. Much more than photography, with its taste for series so conducive to scientific observation, cinema has shifted the demand for objectivity towards a fantasy of the imagination.

Cinema is characterized by its ability to transform every existing thing, human, animal, plant, mineral, into a subject. In a 1903 film, *Electrocuting an Elephant*, a Thomas Edison camera operator filmed the execution of Topsy, the killer elephant of Luna Park Zoo in New York, sentenced to death after causing the deaths of three of her keepers. Her body harnessed, her feet firmly attached to wooden sandals fitted with copper electrodes, the pachyderm stands a few distances from the operator who has positioned himself in front of the crowd that has flocked in large numbers to witness the execution. In her visual history of medicine, historian Lisa Cartwright suggests that this *dispositif* of death, doubled by that of the camera, constitutes a convincing demonstration of the "power of technology to regulate and discipline bodies".[17] Watching this barely one-minute film, however, one is struck by the painful choreography of the animal's body at the moment of death. Something escapes the rigorous organization of the scientific experiment and execution. At the fateful moment, a cloud of smoke rises from the animal's feet. Initially immobile, she wobbles, gently collapsing onto herself as if in an elegant salute, while the smoke modestly wraps her in a white shroud. The violence of the sequence, without bloodshed or sparks flying, lies solely in this almost magical sleight of hand that gives the scene a disturbing singularity, not of a spectacle of attractions but of a moving drama. Chris Marker, Maya Deren and Alexander Hammid, but also Joyce Wieland, have each celebrated this animal singularity, when they became discreet observers of the private lives of their cats[18] or when Marker went wandering through the zoos of Ljubljana and Berlin. The elephant capsizing like a drunken boat in its solitary enclosure (*Slon Tango*, 1990), the huddled rhinoceros with only its ear animated by a sad note, the dazed monkeys and the felines endlessly turning behind the bars of their cages (*Zoo Piece*, 1990), all seem inhabited by an animal melancholy. Not one is a specimen; all are the subjects of their own tragedy.

"To Better Speak of It, Best Not to Speak of It at All" (Disenchantment, Re-enchantment)

Portuguese artists João Maria Gusmão and Pedro Paiva have perhaps best explored this ambivalence of the filmic image, which always shows something more or different than what it was supposed to represent. "How and why we filmed what was never actually seen"[19] they write in an "extra-terrestrial" prologue that strips Karl Popper's falsifiability principle of its value as a means of distinction between science and metaphysics. In a series of short films shot on 16 mm, they use the powers of analogy to test this "theory of exception". The result is a collection of cinematic *mirabilia*, lying on the border between René Daumal's pataphysics and the scientific curiosities of the seventeenth century, which play on the theories of modern science through a comic empiricism. Within this project, animals play a fundamental role. In a series of six shots and over 60 seconds, *Darwin's Apple, Newton's Monkey* (2012) shows gleaming red apples falling slowly under the impassive gaze of stuffed monkeys. These are non-living beings, in a space without coordinates, watching without seeing bodies that fall where they should float. This is the playful empiricism of Gusmão and Paiva: a change of perspective that cinema allows by modifying the experience of space and time, casting a new light on the great scientific discoveries. Newtonian physics is thus thwarted by the speed of images, while the stasis of the taxidermized monkeys questions the principle of Darwinian evolutionism (Fig. 9.3).

How would one tell monkeys about the revolutions of thought produced by Copernicus, Kant or Darwin? "I'm sure monkeys wouldn't be interested in Kant, and Copernicus' theory would be quite irrelevant for Tarzan-like apes. I have, however, a few doubts about natural selection; somehow I can't get over the fact that intelligence is enormously overrated",[20] comments the narrator of the extra-terrestrial prologue. If Copernicus, Kant and Darwin established scientific facts that contributed to the disenchantment of the world, then Gusmão and Paiva's entire *oeuvre*, heirs in this sense to Jean Painlevé, proceeds from a re-enchantment of science by cinema. The film strip, which is their working tool, is not a coquetry in this venture to restore wonder. It upholds a principle of analogy between distinct realities, which is why it is haunted by animal figures: a ray (*Dream of a Ray Fish*, 2011), suspended in the void and traversed by light, both medium and projection surface, an almost immobile turtle (*Turtle*, 2011), a small fish in agony on a plate (*Cowfish*, 2011), or a cassowary perplexed by a landscape painted on a canvas (*Cassowary*, 2010).

Paradises Lost

As we come to the close of this small cinematic bestiary, where we have explored a historical hypothesis according to which animals have appeared on screens at the very moment they began to be driven from their natural habitats, tested the common fate of analog images and animal figures through the experience of time and the motif of the corpse, and inventoried empirical uses of film as a rewriting of so-called "natural" sciences, it is impossible not to wonder what becomes of this story in the age of the sixth mass extinction. The artist duo Neozoon—literally "new species"—examine the relationships between humans and animals in the

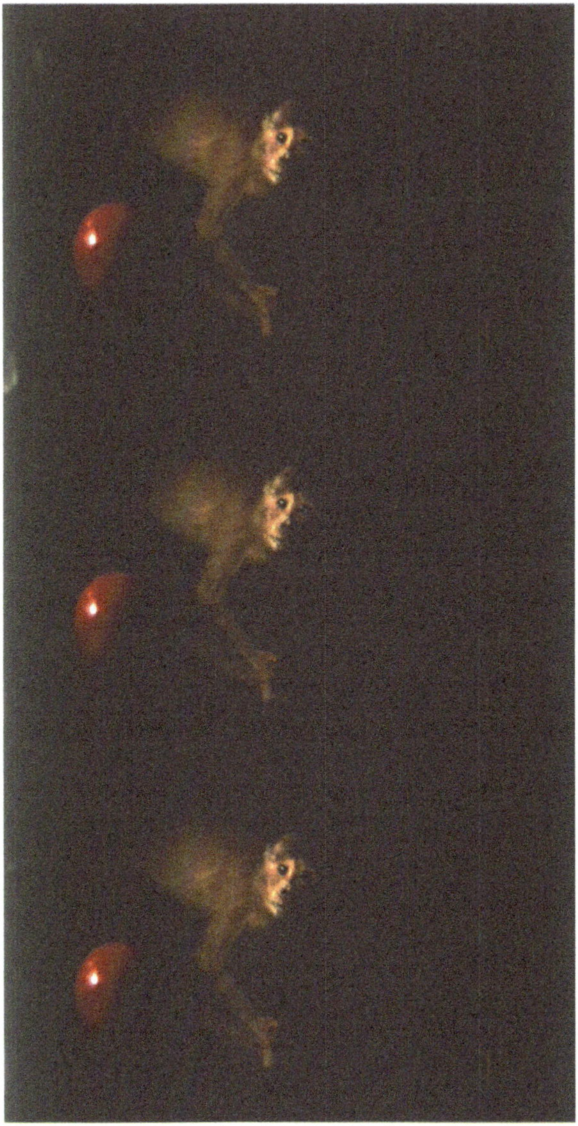

Fig. 9.3 Jose Maria Gusmão & Pedro Paiva, *Darwin's Apple, Newton's Monkey* (2012)

internet age.[21] Their films, made from footage found online, testify to the fact that these relationships no longer obey any direct interaction but depend upon a phantasmagoric imaginary populated by simulacra. In *Unboxing Eden* (2013), American Youtubers film themselves as they carefully open postal packages. The triviality of the gesture soon gives way to the strangeness of the discovery: in these boxes, carefully wrapped in silky fabrics, are snakes deposited like precious merchandise. Of all sizes and colours, the reptiles are carefully extracted from their polystyrene and cardboard nests. Measured, weighed, admired, they have become as luxurious as the animals in Philip K. Dick's dystopian fable. For what interests the artists of Neozoon here is the transplantation of the biblical animal into a disenchanted world, where it is not only tamed but reduced to the status of an object. Ill-suited to this domestic ecosystem, the snake's body undulates between the coffee maker and the sink, like some kind of aberrant simulacrum.

The story has yet to be told of the simultaneous birth of images and synthetic beings, that which links Dolly the Sheep, the first cloned sheep in 1996, to *Babe* the pig-shepherd from Chris Noonan's film (1995).[22] We shall then have to question why both images (since the clone, understood as replica and as simulacrum, is also a "living image"[23]) inaugurate a new ontology of life forms in the twilight of the twentieth century. We will then remember that the fate of animals and images, in the history of analog cinema, was linked by an emulsion bath using animal-derived gelatine, generally produced from the skin or bones of pigs or cattle, and that digital technologies have replaced this photochemical materialism with a computational logic of data and calculation. The digital image no longer preserves the trace of vanished beings but gives rise to the fantasy of a totally artificialized life. Humans and animals no longer inhabit the same garden, each follows its solitary path.[24]

Notes

1. "Buffalo Bill's Wild West was based at Ambrose Park in Brooklyn, New York, from May 12 to October 6, 1894. Numerous performers associated with the show appeared in front of Edison's camera in September and October. Buffalo Bill [William F. Cody] led the first expedition on Monday morning, September 24th, less than two weeks before his organization departed for Europe". Four films about Buffalo Bill and the Sioux warriors were made

on this occasion: *Buffalo Bill, Sioux Ghost Dance, Buffalo Dance, Indian War Council*. In Charles Musser, *Edison Motion Pictures, 1890–1900—An Annotated Filmography*, Washington/Gemona, Smithsonian Institution Press/Giornate del Cinema Muto, 1997, pp. 125–126.

2. John Berger, *About Looking*, New York, Pantheon Books, 1980, p. 1.
3. *Ibid.*, p.19.
4. As Anat Pick explains, these architectures that claim to substitute traditional cages with discreetly enclosed natural spaces, under the pretext of recreating the living conditions of animals in their natural environments, hide another issue, that of creating an immersive setting for the visitor. See Anat Pick, *Creaturely Poetics: Animality and Vulnerability in Literature and Film*, New York, Columbia University Press, 2011, pp. 104–105.
5. Malcolm Le Grice, "Real TIME/SPACE" [1972], in Malcolm Le Grice, *Experimental Cinema in the Digital Age*, London, BFI Publishing, 2001, p. 156.
6. *Ibid.*, p. 157.
7. Michael Lawrence, "Muybridgean Motion/Materialist Film: Malcolm Le Grice's *Berlin Horse*", in Michael Lawrence and Laura McMahon (eds), *Animal Life and the Moving Image*, London, BFI, Palgrave, 2015, p. 73.
8. Akira Mizuta Lippit, *Electric Animal. Toward a Rhetoric of Wildlife*, Minneapolis/London, University of Minnesota Press, 2000, p. 1.
9. *Ibid.*, p. 187.
10. Catherine Russell's remarkable book shows this triangulation between colonialist, ethnographic and pornographic views. See Catherine Russell, *Experimental Ethnography: The Work of Film in the Age of Video*, Durham/London, Duke University Press, 1999.
11. Stan Brakhage: http://lightcone.org/en/film-8557-sirius-remembered.
12. Donato Totaro, "*Stan Brakhage: Death is a Meaningless Word* (Part 1)" Stan Brakhage at the Cinémathèque Québecoise, Montreal, 27–28 January 2001, *Offscreen*, vol. 7, no.2, February 2003. Online: http://offscreen.com/view/stan_brakhage.

13. Brakhage's description as written on the film's page on the Light Cone website, *op. cit.* During the presentation of the film at a retrospective of his work at the Quebec Cinémathèque, the filmmaker explained that the images had been shot as if in a desperate effort to reanimate the dog, to bring him back to life: "*The pictures were photographed as if desperate to try and reanimate the dog, to bring him to life*", *ibid.*
14. Stan Brakhage, http://lightcone.org/en/film-8557-sirius-remembered.
15. Donato Totaro, "*Stan Brakhage*", *op. cit.*, Brakhage quotes Wittgenstein saying: "*Death is a meaningless word because it refers to something that cannot be lived through*", whereas the exact quote would rather be: "*Death is not an event of life. Death is not lived through.*" See Ludwig Wittgenstein, *Tractatus Logico-Philosophicus*, trans. Charles Kay Ogden, London/New York, Kegan Paul, Trench, Trubner & Co. LTD/Harcourt, Brace & Co., 1922, 6.4311, p. 185.
16. Lorraine Daston and Peter Galison, *Objectivity*, New York, Zone Books, 2007, p. 121.
17. Lisa Cartwright, *Screening the Body: Tracing Medicine's Visual Culture*, Minneapolis/London, University of Minnesota Press, 1995, p. 18.
18. Alexander Hammid, Maya Deren, *The Private Life of a Cat* (1946); Joyce Wieland, *Cat Food* (1967) and *Rat Life and Diet in North America* (1968), two films in which Wieland humorously and ingeniously allegorizes the types of domination at work in a capitalist society and the means to resist them.
19. João Maria Gusmão and Pedro Paiva, *Teoria extraterrestre*, Milan, Mousse Publishing, 2014–2015, p. 53.
20. *Ibid.*, p. 183.
21. I thank my student Gaspard Labastie-Coayrehourcq for our conversations about Neozoon's films.
22. Alice Leroy, "Récits des mondes autres qu'humains", *Débordements*, "Terrestres après tout" no.2, Autumn 2020, pp. 107–119.
23. See W.J.T. Mitchell, *What Do Pictures Want? The Lives and Loves of Images*, Chicago, University of Chicago Press, 2004.

24. "Some natural tears they drop'd, but wip'd them soon; / The World was all before them, where to choose / Thir place of rest, and Providence thir guide: / They hand in hand with wandring steps and slow, / Through *Eden* took thir solitarie way" *in* John Milton, *Paradise Lost* [1667], Oxford, Blackwell Publishing, 2007, p. 332 (XII, 645–649).

CHAPTER 10

The Environment, Films and Myself

Rose Lowder

I became a filmmaker by chance. In the early 1960s, when watching artists' films, I noticed that what was perceptible on the screen did not correspond to what was visible when examining the film strip. Wanting to know more, out of simple curiosity and just for myself, I worked for a while to test how the images recorded successively on the film strip overlapped when the film was projected on the screen.

In mainstream cinema, the structure of a film relies on an arrangement of shots intended to produce verisimilitude. To be successful, this effect of reality must be based on precise cinematic conventions. After a few years of experimentation, I was able to see how the relative proximity of images with different characteristics could affect the way in which their visual components would be perceived on the screen. By varying each photogram, I noticed that some images, or certain parts of these images, seemed to dominate or be present on the screen longer than other images. Intrigued, I worked with felt-tip pens and a hole punch on transparent 16mm film strips. I discovered that what we saw on the screen was

R. Lowder (✉)
Lima, Peru
e-mail: roslowder@gmail.com

E. Della Noce and L. Murari (eds.), *Expanded Nature*, Experimental Film and Artists' Moving Image,
https://doi.org/10.1007/978-3-031-70729-2_10

based on the shapes, colours and graphic characteristics of the film strip, according to their relative position on the strip. If I draw a straight line and then punch holes in each frame on the strip, we perceive a circle on the screen without the line ever crossing it. If perforation is only made after two or three frames, a more or less distinct line will appear to be crossing the middle of the circle. By curving the line, it can be made to pass across the screen in front of and then behind the perforated circle. These possibilities rely on the fact that perception is temporally structured in periods of variable length depending on the optical nature of the visual configuration.

Discovering that what could be perceived on the strip did not correspond to what was perceived on the screen (a hole where there is nothing cannot normally become a line) led me to compose films frame by frame in the camera while shooting. The Bolex cameras I use offer many possibilities. They allow you to film alternately frame by frame and at different frame rates with great ease. Also, these cameras can operate in forward and reverse and have frame counters with two discs. One is a metric counter that indicates the length of the exposed film, the other, divided in two, adds up the frames in forward mode and subtracts them in reverse mode from 0 to 50 frames, then totals the frames in forward mode up to 1000 frames and counts down the frames in reverse.

As long as you have a way of making an immediate note of the images recorded, the technical resources of the Bolex make it possible to film images at different points on a reel of film in a given place and at a given time, then to go back to film in another place the parts of the reel still unexposed. Thus, to make *Voiliers et coquelicots* (2001), I filmed the poppies in May in various fields in the Vaucluse intermittently along the film reel, then the boats leaving the Old Port of Marseille in July intermittently on the unexposed parts of the film. As you might expect, the sailboats sail across the poppy fields on the screen, while on the film strip the poppies and the sailboats are on different frames (Figs. 10.1 and 10.2).

Some of the difficulties encountered in making this film are not visible on the screen. First of all, it takes a long time to find poppy fields in full bloom in a country where they are being wiped out by the dominant industrial agriculture, and then I had to deal with the pervasive presence of couples dressed in wedding attire. They always wanted me to stop in the middle of the shoot to take their picture in front of the poppies. Then the weather forced me to go to Marseille three or four times before

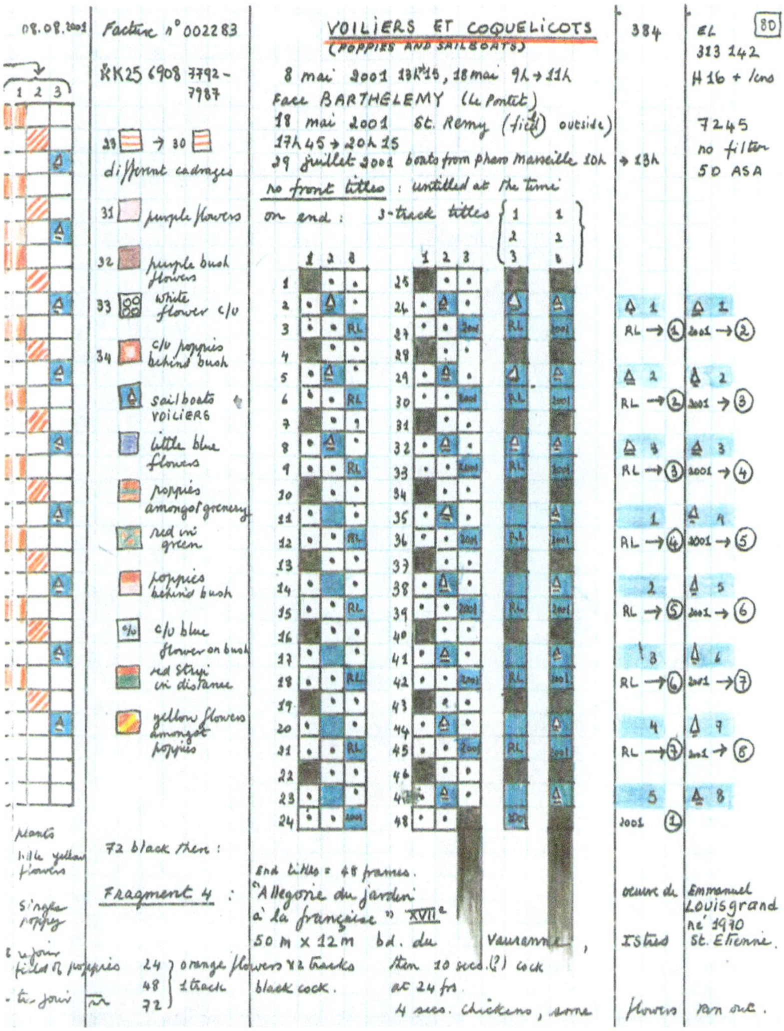

Fig. 10.1 Rose Lowder, Visual Score for *Voiliers et coquelicots* (2001) © All rights reserved by the artist/Courtesy of Light Cone

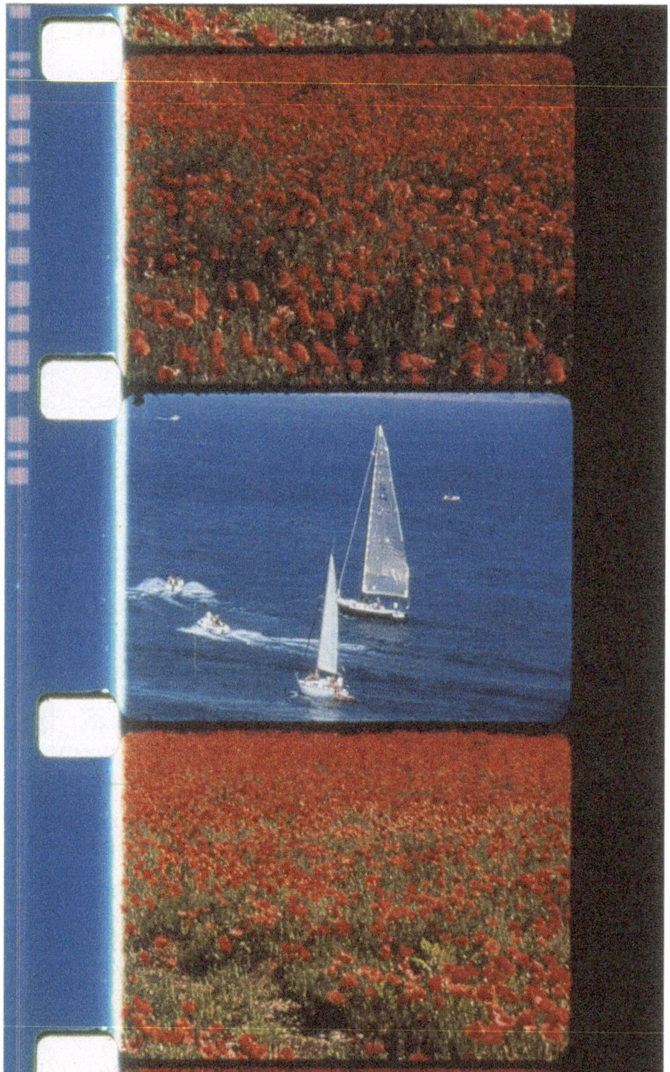

Fig. 10.2 Rose Lowder, Film strip from *Voiliers et coquelicots* (2001) © All rights reserved by the artist/Courtesy of Light Cone

I could film the sailboats. Moreover, the boats are not allowed to unfurl their sails in the harbour, which meant they had to do so relatively far from me. In order to achieve a correct focus, I had to guess in advance exactly where the boats were going to pass through the frame.

Voiliers et coquelicots was a way of cinematically weaving frame by frame two partly unpredictable environments. In other films, it is the movements of the subject, whose particularities are barely perceptible on site, that play a crucial role in their visual appearance on screen.

This is the case in *Roulement, rouerie, aubage* (1978), where the rotational movement of the blades of a paddle wheel catches the sun's rays, obscuring one or two frames of the film each time it passes. The rotation of the wheel's paddles thus inscribes its perceived rolling on the screen. Filmed with a hand-cranked mechanical Bolex, without any camera movement, the aesthetic appearance of the wheel comes from precise focus and framing using black-and-white films of different sensitivities: 7231 with red filter to accentuate contrast, 7222, 7224 and 7247 colour film. The appearance on the screen of the variable concentration of film grain of these different types of film creates a visual poetry of the rotation of the moving wheel's wooden boards.

Roulement, rouerie, aubage was about the rotation of a mechanical object, set in motion by the water of the Sorgue canal, coming from Fontaine-de-Vaucluse. For another film, *Jardin du sel* (2011), I took as my subject organic agriculture closely linked to a primordial component of the environment, the sea. With the agreement of a salt worker, Hervé Gilles, in the salt marshes of Guérande, I was able to focus on the movement of the fine, light crystals of the *fleur de sel*. They float, often in several directions at once, on the surface of the water, over the coarse salt that is at the bottom of the clay salt pond [*oeillet*]. This time I filmed five reels in Kodachrome of about three minutes each, each time continuously, with the camera on a tripod and without any pause or camera movement. I had visited the salt worker's ponds at all hours of the day for several days in a row to see the degree of crystallization of the salt but also to find the ideal place to position myself in relation to the position of the sun. Because of the sun's movement, each reel was filmed from a slightly different location, giving a different version of the subject. Most importantly, I wanted to make visible the nature of the *fleur de sel* crystals set in motion by the wind. With a precise focus set in relation to the framing, my intention was to allow an in-depth perception of the salt grains composing the subject. The use of Kodachrome film, with its particular colouration, helped to underline this perception.

Salt production by an artisan working with sea water is an art and requires constant effort all year long. Often the salt workers help each other with the hard labour of the preparation of the salt ponds. There is an enormous amount of adjustment and meticulous fine-tuning to be done daily, through a carefully designed system of valves, traps and flaps, to control the water level so that the salt can crystallize.

Salt being a vital element for the body, I wanted to show in *Jardin du sel* the ingenuity of a natural element which, in collaboration with human labour that takes into account the quality of the environment, nourishes us. The title of the film recalls that salt marshes were once called "salt gardens" because they were part of agriculture in the same way as vegetable gardens.

The music for the film was composed by François Alexis Degrenier, after having seen the images many times with me. Having music, sound poetry, on images which are visual poetry, makes the work more complex, because each means of expression must simultaneously have its importance in the film and give the work a specific presence on the screen. To finish, I filmed on the remaining Kodachrome reel another contribution to the biodiversity of the environment: wild flowers in a small park in Villeneuve-lès-Avignon, near our house in Avignon.

In *Tartarughe d'Acqua* (2016), the subject was completely unplanned. I was walking through a public garden in the city of Asti in Piedmont, Italy, when, in a very small pond, I found a large quantity of turtles of a species known as *Trachemys scripta* and *Trachemys scripta elegans*, protected by the *Movimento Nationale U.N.A. Onlus Uomo-Natura-Animali*. These turtles had their own habits, linked to the time of day and the position of the sun. The pond was very small for so many turtles. There were also goldfish. In addition to a small fountain, there were three small floating islands. The turtles would swim in the basin for a good while, the larger ones often carrying the smaller ones on their back, then, when towards the end of the morning the sun's rays shone through the large trees and reached the pond, all the turtles would climb onto the small islands to dry themselves. As there was not enough room, they would climb on top of each other, without any aggression, until a small vertical tower made up of three or four turtles capsized as a result of the last arrival, sending everyone into the water.

I was struck by the calm and well-being that reigned in this group of turtles who, once each had found its place, stretched their legs in all directions to dry them off. To film, I had to come several days in a row because the pond only got one or two hours of sunlight a day. Each reel was filmed continuously with a fixed unmoving camera on a tripod. The subject was very unpredictable because on the one hand the movements of the turtles could not be predicted, and on the other the wind would constantly move the small floating islands, which forced me to estimate the focus without being sure that the subject would enter the frame. The only solution was to frame the shot and then, during the filming, to modify the focus slightly but constantly according to the movements of the turtles and the displacement of the islands.

The subject seemed worthy of interest to me because I was sensitive to the fact that a conservation organization had placed these turtles in the public garden for everyone to enjoy. When I was filming in August the park was mostly frequented by low-income families who could not afford to go on holiday. As for the cinematography, I wanted to see how the slow movements of the turtles, on restless islands, could be incorporated into a specifically filmic temporality. As the art of cinema relies on many different kinds of temporalities, it is difficult to bring them together in a satisfactory whole. There is the time during which the film is made, the time of the subject being filmed, the time of the length of the film projection, and, most importantly, the time of the work itself: the time perceived or felt in front of the screen. Apart from a few exceptions, as these four temporalities are not in the same order of ideas, they do not coincide. This can also happen in music, as Igor Stravinsky explained: "I know there are parts of Agon which, for the same chronometric duration, contain three times as much music".[1]

In the four films mentioned here, what is felt when they are projected depends as much on the energy of the filmed subjects, leading to associations of ideas, as on the way they were recorded. Seeing boats sailing among poppies cannot be perceived in the same way as a mechanical wheel which through its movement allows the audience to perceive and feel the operation of its essential components. The close-up image of floating *fleur de sel* grains, moved by the wind at varying speeds, leads us to enter their environment in a more meditative way. Nevertheless, these films share a common goal: to seek ways of translating the filmed elements into singular presences on the screen.

Achieving exceptional acuity in an artistic work is an incredible difficult task, regardless of the art in question. It depends on too many factors for it to be a matter of technical virtuosity. As Francis Marmande said, speaking of the double bassist Arvell Shaw: "For a long time, and even during his lifetime, all bassists without exception played 'better' than the amazing Arvell Shaw. But none 'as well' …."[2]

I still remember the dance performance by Merce Cunningham in the Cour d'Honneur of the Palais des Papes in Avignon, when, crossing the stage from right to left, amidst the young and excellent members of his troupe, the master himself, despite his age and the arthritis in his toes, performed some unforgettable dance steps. Similarly, towards the end of her performance at the Centre Georges Pompidou in Paris, among dancers of different shapes and sizes, each performing according to their morphology (even one of the stage managers played a role while continuing their work), Yvonne Rainer crossed the stage with some equally unforgettable dance steps.

It becomes very difficult to describe these moments of exceptional sensibility which in every case are the result of the work of a lifetime. In the noblest sense, art and science converge because in both the means must be sparing in relation to the achieved end. These are the terms in which Jacques Monod (Nobel Prize for Medicine in 1965) spoke of "elegance in science" because "if a theory is unwieldy and unattractive, it is false".[3]

Today the arts, sciences and traditions sometimes converge on common ground. The author Marie Hélène Lafon evokes the tradition of cowbells in her native Cantal. Not all cows were given bells, only some, which meant that the overall sound produced a musical quality. The farmer's own art met another art: music.[4]

In all these cases, through music, dance or cowbells, you can strongly feel the radiance of artistic presence. Nourished by such examples, I try to compose the image in such a way that it allows the viewer to distance themselves enough from the chosen subject to be able, by deepening their perception of the film, to see or appreciate it differently. I try to work between a figurative image and the cinematic possibilities modulating the elements of the filmed reality. I thus try to go deeper into the subject by moving away from a completely figurative or totally abstract image.

I was born in Miraflores, in the province of Lima in Peru, and I lived there all my childhood and adolescence. During that time, I attended many outdoor painting classes and made several trips through the Andes, equipped with everything I needed to paint various landscapes. The reason I use the environment as the subject of my films is because very early on, as I travelled the paths of the farmers of the Andean cordillera, I realized the importance of our relationship with the environment. The way of life of these Andean communities is based on an essential idea: to form with nature a single whole, as opposed to the current conception that thinks that humans are above nature, which gives them the right to dispose of it at will. It is only after making nature a resource to be exploited to exhaustion, causing numerous disasters, that we begin to perceive the importance of the unity of living beings, nature and culture.

Despite much greenwashing rhetoric from those in power, the decisions made favour an unsustainable system where short-term economic interests take precedence over the preservation of ecological biodiversity and the environment, which are closely linked to the health and well-being of populations. In agriculture, for example, only the size of farms is taken into account, not their farming methods. This is how industrial agriculture, which consumes large quantities of water, agricultural inputs and biocides, receives priority financial support, to the detriment of versatile family farming which respects the environment. The environmental and health costs associated with this mode of production are never taken into account: the inevitable impoverishment of humus in the soil due to the massive use of synthetic nitrogen fertilizers and pesticides on which cereal monoculture without livestock depends; the pollution caused by intensive livestock farming; and the elimination of bees, birds and butterflies through the use of neonicotinoids. As a result of this widespread use of chemical products, plants are less resistant against pest insects, which are fond of excess synthetic nitrogen, leading to a growing use of polluting products that destroy biodiversity and cause a considerable increase in cancers, Parkinson's and Alzheimer's diseases and many other civilization ailments. And yet, other traditional farming methods exist, such as crop rotation, including legumes that produce natural nitrogen.

Despite the research of hundreds of scientists, specialists and renowned agronomists proving the extent of the disaster, public authorities ignore these whistleblowers, who are continually harassed and legally and economically threatened by the chemical industry lobbies, who use all sorts of manipulations to conceal the known harmfulness of their lucrative products.

Given the destruction of biodiversity, air pollution and the destruction of the ocean, it seems only natural that the world should be in the state of climate disruption in which it finds itself today. And it is not the so-called "ecological" choices, such as, for example, the electric car, that are the answer to the problem. The manufacture of these cars is very energy-intensive, as is the manufacture of their batteries which is closely dependent on the mining industry for the extraction of rare metals—lithium, cobalt, nickel and manganese—coming from countries like Chile, the Democratic Republic of Congo, New Caledonia or Gabon. These are industries that often employ children in appalling working conditions, so as well as polluting the environment they destroy the health of local populations. We should also bear in mind that electricity in France comes mainly from nuclear power, also presented as a "clean" energy, whereas its production depends on uranium mined in distant countries to the detriment of the environmental and health conditions of their inhabitants. Moreover, there is no solution for disposing of radioactive waste, and nuclear accidents are making part of the world uninhabitable.[5]

The fact that the health of life on earth is directly linked to the biodiversity of the environment is one of the first reasons that led me to develop my film work with nature. Contact with nature also presented me with a considerable artistic challenge: the unpredictability of the subject. Thus, I moved away from description and narration so that they could not absorb attention at the expense of the sensorial qualities of the image. My intention is to give importance to the imagination sparked by the poetry of cinematic qualities and to draw attention to the all-important subject of nature.

Notes

1. Igor Stravinsky, Programme Catalogue, Cité de la musique, 2012–2013. Unless otherwise noted, all translations from original French are by the author and translator.
2. Francis Marmande, "L'album qui m'a fait aimer… La contrebasse", *Le Monde*, Friday 28, August 2020. Arvell Shaw was notably the bassist for Louis Armstrong.
3. Jacques Monod, Interview, *Les trésors de Radio France*, broadcast on 20–21 January 2021.
4. Marie Hélène Lafon, "Musique et moi", *France Musique*, 17 January 2020.
5. See, on these environmental themes: Vandana Shiva, *Who Really Feeds The World?*, London, Zed Books, 2015; Marie-Monique

Robin, *La fabrique des pandémies: préserver la biodiversité, un impératif pour la santé humaine*, Paris, La Découverte, 2021; Claude Aubert, *Les Apprentis sorciers de l'azote: la face cachée des engrais chimiques*, Mens, Éditions Terre vivante, 2021; Pierre Péguin, Yvo Rens, "Voiture électrique: décryptage d'un engouement", *Nature & Progrès*, n°132, April–May 2021; and *L'Association Bloom*, c/o Oxfam France, Paris, which, after years of struggle with other NGOs, has obtained the total and definitive ban on electric fishing by European institutions.

CHAPTER 11

Taking Time

Chris Welsby

In a recent broadcast of CBC Radio's 'Unreserved,' I listened to Rosanna Deerchild interviewing a number of First Nations office workers in Toronto. One of the people interviewed had a habit of going across the road to a small park on his lunch break, rather than heading to the local cafe with his co-workers. As it turned out, he didn't take a book, or a pen and paper, or a cell phone. He didn't even take a sandwich or a cup of coffee; he just sat quietly for an hour in the one little patch of green that was available in his surroundings. When asked what it was that he was doing for all that time, he replied, "Oh, just staying in touch". As if it were the most natural thing in the world.

As I begin this chapter, it is August of 2020. The world is in the grip of a global pandemic and the catastrophic effects of climate change are everywhere in plain view. Half of the West Coast of North America is burning as I write. In my home province of British Columbia, wildfire casualties compete with daily death tolls from COVID-19 for attention from the ad-choked media.

C. Welsby (✉)
British Film Institute, London, England
e-mail: welsbysfu@gmail.com

E. Della Noce and L. Murari (eds.), *Expanded Nature*, Experimental Film and Artists' Moving Image,
https://doi.org/10.1007/978-3-031-70729-2_11

To suggest that our planet is trying to communicate something would be an understatement. And its tiny toxic messengers have no need of social media: in my neighbourhood, the air is thick with smoke and loaded with poisonous particulates, too fine for a facemask to filter. They enter the bloodstream with the ease of a radio signal, announcing, perhaps, that it is time to look up from our screens and engage more productively with our endangered world. My intention in my art practice over the last half-century has been increasingly to foster the kind of attention that will promote and support such engagement.

It is no secret that the arts have become increasingly marginalized and experimental film even more removed than ever from the general public. However, chaos theory maintains that a little bit of input can produce an enormous change in output, so perhaps we should not despair too quickly.

We are challenged by these calamitous times to expand our limited sense of who we are. It is not a great leap to surmise that we would all find it harder to tolerate environmental destruction if we did not feel so separate from the environment. If we were able to experience ourselves as part of nature—not just cerebrally, but in an embodied way—we might be more committed to saving it. And by extension, saving ourselves.

"If" is the operative word here. It is no mean task to shift into a more embodied experience of ourselves and nature. Will future advances in technology support such a shift? I have my doubts that it will be effected through the transhumanist vision of a technologically enhanced intellect and physiology. There is no doubt that new technologies will play a major role in our future as a species, but there is a real danger that, unconsciously or not, they will continue to serve the same corporate elite and perpetuate the human predilection for domination and control.

> Technology is not neutral. We're inside of what we make, and it's inside of us. We're living in a world of connections—and it matters which ones get made and unmade.[1]

Regarding AI, it would be prudent to ask what ***kind*** of intelligence we are hoping to enhance. Will it perpetuate the same old dualistic split between humans and nature or will it support a more participatory engagement with nature? Will it reflect the wisdom and intelligence of our Indigenous cultures?

In what follows, I will briefly discuss the early influence of painting on my moving image practice and I will describe three pivotal experiences of landscape that I have not documented before: an encounter with nature in the New Forest, UK, when I was younger; a more recent study of the pond in my front garden; and my lifelong affair with sailing. I will reflect on my art practice in light of the current environmental crisis, with a focus on three interrelated themes: my own embodied experiences of the natural world; some of the writings that have helped me to realize and develop my ideas; and a few examples of my art projects, past and present. The projects will include two early films shot in the UK: *Seven Days* (1974) and *Park Film* (1972); two New Media projects made in Canada: *Time After* (2010) and *Tree Studies* (2006)*;* and two recent films shot in Mexico: *Desert Spring* (2017) and *Casting Light* (2019).

My continuing challenge is to generate a collaborative experience in which technology is not invisible and not being used to dominate the landscape—or re-purpose it as a backdrop to some kind of human drama. Instead, authorial control is intentionally reduced to allow nature to become a driving force in a partnership where agency is shared with technology (the camera), the viewer and the artist. My hope is to evoke a shift in the habitual human experience of being a separate subject in a world of objects: to invite the viewer to a more inclusive experience in which the sense of separation has a chance, if briefly, to dissolve.

Participating Consciousness and the New Forest

Cultural historian Morris Berman adopted the term "participating consciousness" in 1981 to describe an ancient mode of human cognition in which the perceiver and the world perceived are not experienced as separate.[2]

In August of 1970 I had a seminal experience of participating consciousness that has sustained and informed my art practice ever since.

It was one of those blustery days in the New Forest when one weather system after another keeps rolling in off the Atlantic. Heavy showers were followed by moments of intense sunlight and saturated colour. Huge black-and-white clouds drifted ominously over a cobalt blue sky. I was on horseback, and as I rode through stretches of open heathland and densely wooded hills, the sobering constraints of my city life fell away and the sheer intensity of each passing moment made me laugh aloud.

I came to a clearing in a grove of ancient oaks. The grass was glistening and light was reflecting off a million raindrops. In the clearing, a pool of silver-and-black water was pitted with rain drops and animated by fierce gusts of wind. Dark cats-paws competed for the surface, fragmenting the reflections of passing clouds and swirling branches. For what felt like an age, the horse and I were motionless in the clearing, rain pouring over us in one moment and bright sunlight steaming our coats in the next. The whole scene was alive, suffused with a supernatural light. Wind rushed through the clearing like a dancing dervish, casting leaves and branches before it in a wild cosmic dance.

I was transfixed, no longer separate from my surroundings. It was as if everything, myself included, was connected to everything else by an invisible web, finer even than the glistening webs of spiders that hung in the trees.

Seven Days

In *Seven Days*, I tried to recapture the essence of this experience through film in a very different landscape on the west coast of Wales. The shape and duration of the film are co-authored by three forces: my actions as the camera operator, the rotation of the Earth and the rapidly changing patterns of the weather.

The seven days were shot consecutively and each day started at the time of local sunrise and ended at the time of local sunset. One frame was taken every ten seconds throughout the hours of daylight. The camera was mounted on an equatorial stand—a device used by astronomers to track the stars. In order to remain stationary in relation to the star field, the device was aligned with the Earth's axis and rotated on its own axis every 24 hours. Rotating at the same speed as the Earth, the camera was always pointing at either its own shadow or the sun. Image selection (sky or Earth; sun or shadow) was controlled by cloud cover—whether the sun was in or out. If the sun was out, the camera was turned toward its own shadow; if it was in, the camera was turned toward the sun (Fig. 11.1).

A directional microphone was used to sample sound every two hours and the samples were later cut to correspond in space and time with the image on the screen.

The overall shape and pacing of the film were created through the interactions of the filmmaker, the rotation of the planet, the film technology (i.e., the camera) and the ever-changing weather.

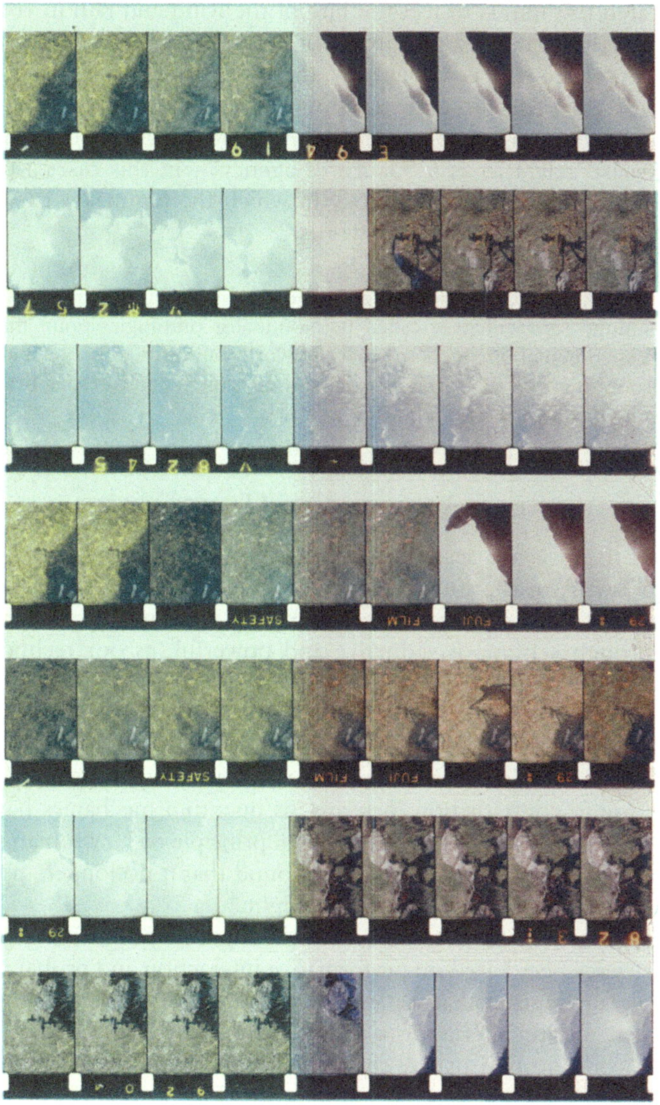

Fig. 11.1 Chris Welsby, *Seven Days* (1974) © All rights reserved by the artist/ Courtesy of Light Cone

To watch this film is to feel the bite of the wind and rain in your face, along with the heat from the projector lamp, the chatter of the shutter, the chafing of the film as it rolls onto the take-up spool and the projector beam as it casts the flickering shadows of the camera onto the movie screen. My intention was to create what Andrew Pickering, social historian and physicist, refers to as a "dance of agency"—in this case, a partnership between nature and technology in which the filmmaker participates without directing. In Pickering's words:

> Dualist understandings foster a worldly stance of domination and "enframing," as Heidegger called it. Nondualist ontologies point instead to an experimental openness to what the world can offer us, a stance much less prone to evoke the sorts of social and environmental catastrophes that are coming to characterize the 21st century.[3]

Shared Agency and Homeostasis

The idea that we are separate subjects in a world of separate objects emerged as part of the Enlightenment and strongly informed the reductionist sciences that followed. The belief that we can objectively see the world through a frame is seductive and powerful, as our technological progress of the last few hundred years will attest. But this limited understanding of the world ran into serious problems in the post-relativistic twentieth century.

To believe that the mind has an objective existence outside of nature is, of course, a contradiction in terms, as quantum mechanics has made clear. By the mid-1920s, this fundamental principle of Newtonian science was brought into question when it was found that it no longer applied to events as they approached the speed of light.[4]

With the arrival of chaos theory and a proliferation of new cross-disciplinary sciences based on performative interaction and probability, we began to really register that we are not separate from our surroundings—that in fact, we are all, scientists included, engaged in a dance of agency with the world around us. By the second half of the twentieth century, these ideas migrated to the humanities, kindling a renewed interest in Indigenous cultures and Eastern cosmologies.

The writings of Gregory Bateson first interested me in the phenomenon of shared agency as expressed by homeostatic systems.[5] The term 'homeostasis' is used to describe self-regulating systems, natural or man-made, that are able to maintain a stable internal equilibrium despite fluctuations in their immediate surroundings. A good example is the human body, which is able to maintain a temperature of 98.6° Fahrenheit amid wide variations in external temperatures. Bateson's *Steps to an Ecology of Mind* was a strong early influence on my practice. In particular, the notion of shared agency in self-regulating systems has influenced the construction of many of my movies and installations, including *Seven Days*, *Park Film* and *Tree Studies*.

THE POND

An influence that continues to delight me and teach me about self-regulating systems is the pond in my front garden. It grows a filigreed layer of reflective weed during the hot, rainless summer that protects it from evaporation. As clouds gather through the fall, rain soon tops up the water level; as winter arrives, the protective layer of weed thins out and becomes compost. Slime quickly follows, and this rich food source attracts bugs and worms. These are now eaten by two fat koi who further process the nutrients and encourage the water lilies to grow. The lilies appear just in time to shelter a stampede of mating frogs that emerge noisily from the nearby woods in the spring, leaving a trail of eggs on everything that grows. A heron shows up from time to time, hoping to catch the fish or frogs that become more visible as the water loses its protective layer of weed. And so it goes, year after year—a self-sustaining, homeostatic ecosystem that draws its sustenance from its immediate surroundings, providing habitat for fish and frogs, and visual pleasure for me.

I am no biologist, but even I can see that this pond is not operating in isolation or on blind chance alone. The various phases of its yearly cycle are a lovely example of homeostasis in nature. Each life form shares agency with the others in the maintenance of the ecosystem as it responds and adapts to changes in its environment—evidence, perhaps, that there are forms of 'cognition' beyond those we have traditionally ascribed exclusively to humans.

In fact, beginning with the Santiago theory of cognition, there is a growing body of research that maintains that evolution itself is a cognitive process.[6]

PARK FILM

I used the idea of a homeostatic system in *Park Film*. Recorded over a period of three consecutive days, the pacing of the film is determined by the flow of people along a busy pathway in a London park. One frame of film is exposed as a person on the pathway enters the frame and a second frame is exposed as they leave. (The viewer can clearly see the build-up of superimposed figures at each side of the image.) The flow of people through the park is in turn determined by morning and evening rush hours, the weather and chance. On a stormy day, for example, walking home through the park will be considerably less attractive than catching a bus, so pedestrian traffic will likely decrease. Less predictably, a broken traffic light between Notting Hill and Knightsbridge may abruptly stop or wildly increase the flow of pedestrians, depending on weather and time.

The number of film frames exposed and the interval between each will vary accordingly, and through all of these moment-by-moment interactions, the speed of the action we see on the screen will slow or accelerate.

It is the shared agency between the weather, the city, the movement of people and the mechanics of the camera that give this film its unusual pacing, its duration and its dramatic contrasts of light and weather. To quote Andrew Pickering once again:

> I think it makes sense to say that everything in the world—cats and dogs, rocks and stones—are all engaged in performative interactions with their environment. ...Dances of agency have their own inner dynamics and an emergent quality—in dances of agency we find out how to react to the unexpected; their trajectory takes shape in real time, rather than being structured by pre-existing causes...[7]

SAILING

When we travel by car we are well protected from the weather. It is as if we were watching it on TV. We pass through it, but we are not connected to it. In a sailboat, however, we are actually *in* the weather. We have to work with the weather to get from A to B. It is a collaboration between human, technology and nature.

In a car, you can be in a hurry. You can drive and still stay in your head for most of your journey. In a sailboat, you have to be in your body and in the moment, with all of your senses attuned to every nuance of wind and tide. The journey really is the destination, and your participation is a partnership. You are sharing agency with nature and technology.

At school I was lousy at football and cricket, so they put me in sailing boot camp in hopes that I would sail for the school. As part of our training we were expected to be able to sail blindfold to windward. Relying only on our senses, we learned to assess the full power of the wind in the sails, the currents on the water, and the responses of the boat to every fluctuation in direction and speed.

MY FATHER

That said, one doesn't need a sailboat to make a connection to the natural world. My father, who was a meteorologist in the RAF, taught me that if you stand with your back to the wind in the northern hemisphere, high pressure is always to your right. If you slowly rotate, keeping your back to the wind as it shifts, you will always have some idea of what the weather is likely to do next. Thus, using nothing more than the hairs on the back of your neck, you can be connected to the gigantic thermal forces that drive our weather and climate. With one simple exercise, you are no longer a subject in a world of objects, but a participating element in a dance of agency where collaboration takes the place of domination and control.

MY FIRST FILM

In the late summer of 1969, I returned from a keelboat race between the south coast of the UK and Vigo, in northern Spain. It had been a tough race with gale force winds for much of the way. We had staggered into Poole harbour just before dark the night before and I was sitting in the warmth of the sun in a state of exhausted reverie, watching as the boat swung safely on her mooring. On a whim, I borrowed the skipper's super-8 mm camera, tied it to the wind vane on the self-steering gear and let the wind and tide make a film. This was the first of many wind-powered projects to come.

INSTALLATIONS

Between 1995 and 2015, I enjoyed a long and productive collaboration with Vancouver artist and software designer, Brady Marks. Together we made a number of large-scale computer-driven gallery installations. Three of them, *Trees in Winter* (2006), *Tree Studies* (2006) and *Heaven's Breath* (2009) were driven by real-time changes in wind speed and direction as they were happening outside the gallery. A fourth installation, *Time After*, is neither a movie nor a still, but a technological hybrid designed to record the larger timescales of natural events like day and night, seasonal changes and even climate change.

For me, the promise at the heart of digital technology was that it would enable me to present current real-time events in the gallery—rather than limiting me to filmed records of retrospective events (e.g., *Wind Vane* (1972) and *Seven Days*).

The new technologies opened up a wide range of creative possibilities. I have chosen *Tree Studies* and *Taking Time* to represent this period of my art practice.

TREE STUDIES

Tree Studies was a weather-driven gallery installation commissioned by the Gwangju Biennale, South Korea, in 2008. This installation is one of a series of similar works with the generic name *Trees in Winter* (2006), borrowed from the poem of the same name by Silvia Plath.

The interactive system that drives the installation takes data in real time from wind sensors positioned on four different continents around the globe. The data is relayed to the gallery in real time via the internet to determine frame rate, image and sound edits, and sound mix levels. In effect, the installation operates like a wind-powered editing suite in which all decisions are made by incoming weather data (Fig. 11.2).

The incoming data is displayed on a console on the gallery wall, prominently exhibited as part of the installation. One of three different viewpoints of the same tree is visible to the viewer at a time, depending on wind speed and direction. Covering the circular floor of the gallery, a large compass rose made of tiny LED lights indicates the wind direction at the weather station that is driving the work in that moment.

Input from the weather station pushes the system toward instability, cross-cutting from one camera angle to another. Shifts between modes of generative sonic representation constantly create new and unexpected combinations of image and sound from the same set of pre-recorded components. The 'shape' of the work at any given moment is governed by the forces of nature, both local and global. The installation is a

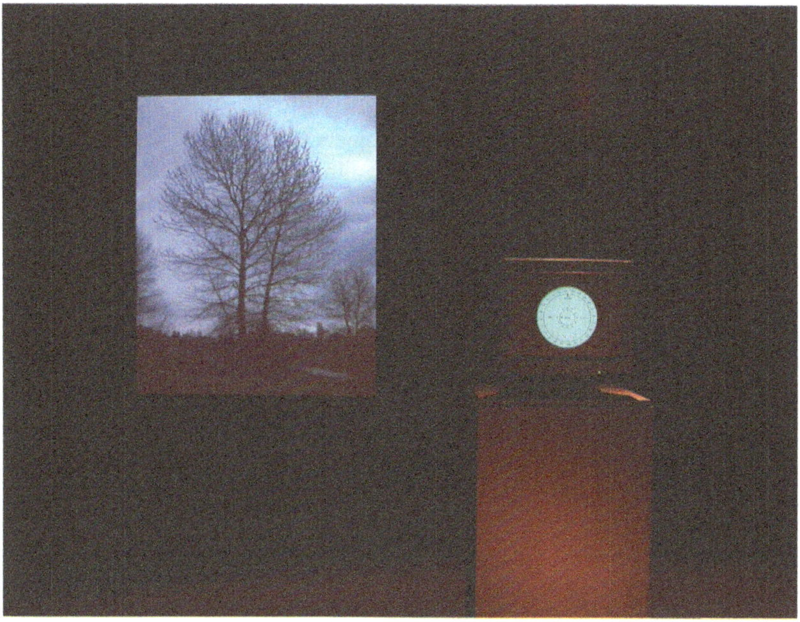

Fig. 11.2 Chris Welsby, *Trees in Winter* (2006) © All rights reserved by the artist/Courtesy of Light Cone

collaboration between nature and technology in which both are seen as interconnected parts of a living system.

As with actual trees in an actual landscape, the representation in the gallery changes its form and appearance in response to input from the weather. The flickering, ephemeral nature of the projected image combines with the changing winter light to create an uneasy equilibrium between the power and presence of the tree, the transitory nature of the light and clouds, and the human presence in the landscape. The overall feeling of the work reflects both the enormous thermal energy that drives global weather and the vulnerability and transitory nature of all living systems.

Painting and Film

I was a painter before I became a filmmaker, and my experience with painting was an important influence in my early development as an artist. I still see it reflected in the kinds of projects I choose to make.

Both the landscape painter and the filmmaker are inevitably dealing with the problem of time. For the painter, very briefly, I propose that the process works something like this: The painter observes a breaking wave, mixes a little grey paint, then puts a dab on the canvas. By this time, the wave has gone. The painter then sees a little red boat, mixes the colour and places it on the canvas. The sun then comes out, the boat moves on and everything looks different. Another wave breaks and this is added to the shoreline, close to the first wave, just as a walking figure appears on the beach. The painter mixes the paint to correspond with the blue of her coat and lays it on the canvas—by which time she is gone… And so it goes, until the canvas is filled. The end result looks very much like a still image, but is actually made up of multiple moments in time.

I became fascinated with the realization that the daubs of paint were much like single frames in a time-lapse film. As the realization became more persistent, the focus of my art practice shifted from painting to film.

TIME AFTER

In this installation, a custom software program slows the digital imaging process down enormously. The idea was to create a camera system capable of recording the long slow changes normally associated with the extended durations of natural change that are so often imperceptible to the human eye (E.g., in weather—seasons, tides, daylight, etc.). As the refresh rate gets slower, human activity becomes less noticeable, eventually disappearing altogether.

Instead of using the lines and frames of a conventional video image, the software chooses small groups of pixels at random from the web-cam image and relays them to the image in the gallery. The process is continually repeated, gradually filling out the image as our painter in my earlier example filled the canvas. Unlike the painter's work, however, the image is never complete because the pixels are constantly erasing and replacing themselves. Using this system, the photographic image is never blurred by motion, as it would be by its photographic relative, the 'time exposure'.

The whole image is always sharp, no matter how fast or slow the moving object, event or sequence of events may have been.

The project was viewable live online in Nantes for three years (14 June 2013–14 June 2016). I have a rather ambitious plan to make a 100-year version of this piece.

> Time present and time past / Are both perhaps present in time future/ And time future contained in time past / If all time is eternally present / All time is unredeemable. T.S. Eliot, first of the *Four Quartets.*[8]

Recent Work

In my recent art practice I have chosen to take a break from the large-scale computer-driven installations of the previous fifteen years and return to working alone with a single portable HD camera. My practice now is simpler, more spontaneous and less time-consuming. The change coincided with my departure from academic life and the opportunity to live half-time in Mexico where I have experienced some very different landscapes.

DESERT SPRING

Shot in the Sonora Desert, Mexico, in March of 2017, I view this video as a joyful little musical trio composed of camera (autofocus), wind and some remarkable desert flowers that I came upon high on the shady side of a mountain. There is hope in this movie—prompted in part by the surprise of finding such delicate beauty in the midst of such harsh climatic conditions.

The view of the desert from high on the mountaintop was spectacular that day, but I wanted to capture the close-up detail, to draw the viewer's attention away from the more dramatic long view to what might easily have been overlooked in the face of it.

I have generally avoided the picturesque and the long shot in favour of the close-up. The media is already choked with spectacular images, historically of beauty and more recently of disaster. We watch the ice caps melting, the plastic islands growing, the air becoming more poisonous, and it fills us with despair. I think of this as the New Sublime—disaster porn that raises TV ratings and sells products. It is fast-paced, well-shot and sexy, but it doesn't inspire change, any more than the warning images on a cigarette pack inspire the addict to quit. Despair is the problem, not the solution.

In *Desert Spring*, I invite the viewer to stop for a moment, as I did on my walk, and take time to look at these exquisite little plants clinging to existence by the side of a dusty desert trail. I want to offer an experience of a nature that is alive and well and still accessible. I want to work with images that are similar to the ones most people can choose to experience: in their local park or back yard, on their day at the beach or their walk in the woods.

> To live in the present moment is a miracle. The miracle is not to walk on water. The miracle is to walk on the green earth in the present moment... Thich Nhat Hanh, *The Miracle of Mindfulness*.[9]

CASTING LIGHT (MEXICO, FEBRUARY 2019)

Casting Light is my most recent video, shot just before the pandemic travel bans of early 2020. It is a comparatively late addition to the long and respectable tradition of the single-take film. The success of this genre generally depends on the filmmaker's ability to isolate some aspect of the world and 're-present' it with a minimum of mediation. It is my hope that in choosing the single-take approach, I might facilitate a moment of "participatory consciousness"—i.e., a shift in perception in which the gap between subject and object is dramatically reduced. I'd like to think that the more immediacy we are able to bring to our experience of the world, the greater our chances might be of discovering what is needed to save it.

The video is a low-angle, continuous take of a clear pool and waterfall deep in the tropical jungle. An ape-like rock formation standing ominously at the crest of the falls is playfully referred to as an ancient river god by the voiceover that dominates the first few minutes of footage. Fish dart busily back and forth in the pool, unaware that they are probably doomed. The river downstream is already dry and it will be months before the rains come—if they come at all—to replenish this once fertile river valley. Oblivious to their fate, the fish continue to compete for food and shade—crude, if obvious metaphors for the human predicament.

When the voiceover falls silent, viewers are freed to immerse themselves in the clear water and to contemplate the scene with fresh immediacy.

Drawn almost hypnotically into the magical world below the surface, they can now bring their own thoughts, associations and responses to the mix, sharing agency with the filmmaker and creating a parallel narrative of their own.

Relevant Film Clips

https://chriswelsby.uk
https://chriswelsby.uk/film/park-film-preview/
https://chriswelsby.uk/film/seven-days-preview/
https://chriswelsby.uk/installation/taking-time/
https://chriswelsby.uk/installation/trees-in-winter/
https://chriswelsby.uk/film/desert-spring/
https://chriswelsby.uk/film/209 *(Momentum)*.

Notes

1. Donna Haraway, "You are Cyborg", *Wired Magazine*, February 1997, accessible online at: https://www.wired.com/1997/02/ffharaway/.
2. The separation of the observer from nature is one of Western science's most cherished principles and the Christian belief in the immortal soul as separate from the body would appear to reinforce this position. However, Aboriginal cultures and our own pre-Christian past reveal that this was not always so and, in fact, our current Cartesian ontology is only a few hundred years old. Participatory consciousness involves making a direct and embodied experience of one's tools and surroundings, and here I apply this to both the idea and the experience of sailing a small boat blindfold, where every nuance of wind and water must be taken into account I also apply it to the history of my art practice, which can be characterized as a search for ways to translate these ideas and sensations into a cinemagraphic experience, and perhaps make way for a more inclusive experience of the natural world.
3. Andrew Pickering, *New Ontologies—Sketches of Another Future: Cybernetics in Britain, 1940–2000*, Chicago/London, The University of Chicago Press, 2008, p. 392.
4. "In the strange field of quantum mechanics, the chance-like behavior of subatomic matter turned out to be extremely resistant to

reductionist methodology. In particular, it was the question of 'decoherence' that was vexing to those who insisted that matter should behave consistently as a particle or as a wave. When an electron is fired at a metal plate with two slits in it, the electron passes through either one or both slits as if it were both a particle and a wave and … worse still, this distinctly non-Euclidian behavior is effected by the observer", Chris Welsby, "Expanded Cinema: Notes on twentieth-century encounters with art, science and technology", *MIRAJ*, vol. 5, issue 1 & 2, pp. 88–107.

5. Gregory Bateson, *Steps to an Ecology of Mind*, Boulder, Paladin Books, 1973.
6. Humberto Maturana and Francisco Varela, *Autopoiesis and Cognition: The Realization of the Living*. Dordrecht, D. Reidel Publishing Company, 1980.
7. Andrew Pickering, "Against Human Exceptionalism", written for the workshop *What Does it Mean to Be Human?*, Exeter University, Department of Sociology and Philosophy, 25 January 2008. See also: Andrew Pickering, "Brains, Selves and Spirituality in the History of Cybernetics", Berlin, Max Planck Institute for the History of Science, November 2007, accessible online: https://metanexus.net/h-brains-selves-and-spirituality-history-cybernetics/.
8. T.S. Eliot, *Four Quartets*, New York, Harcourt, 1943.
9. Thich Nhat Hanh, *The Miracle of Mindfulness*, Boston, Beacon Press, 1999.

CHAPTER 12

The Nebulous Aesthetics in Philippe Cote's *L'Angle du monde* and David Dudouit's *Île de Sein*

Yaniv Touati

Following the footsteps of Jean Epstein's insular films, *L'Angle du monde* (2006) by Philippe Cote and *Île de Sein* (2009) by David Dudouit offer a documentary and poetic take on the Breton islands.

Located in Ouessant, Sein and Molène, three islands on the edge of the Iroise Sea where Epstein shot his film *Mor'vran* (1930), *L'Angle du monde* consists of a sequence dedicated first to variations of the sky, then to images of the sea, seen mainly through a boat trip. At the end, views from the shore, both in black and white and in colour, finally show us the meeting of the water and the clouds.

More anthropological than *L'Angle du monde*, *Île de Sein* presents not only nature (tide cycles, course of the sun and clouds, undulation of seaweed, etc.) but also the daily life of the inhabitants on this Finistère

Y. Touati (✉)
University, Paul Valéry Montpellier 3, Montpellier, France
e-mail: yanivtouati@gmail.com

E. Della Noce and L. Murari (eds.), *Expanded Nature*, Experimental Film and Artists' Moving Image,
https://doi.org/10.1007/978-3-031-70729-2_12

island, shadowed by harsh climatic conditions. Through black-and-white shots that are sped up by a frame-by-frame recording, or, more rarely, slowed down, the film describes in detail the movements of fishing and kelp-harvesting vessels, the gestures of the kelp harvesters, and even the walk of passers-by. Having passed away prematurely in April 2015, David Dudouit was only able to show the public the images that he had just started to sort and assemble. His friends and colleagues at the L'Abominable laboratory[1] then proceeded to finish the editing work whilst respecting the unity of the reels left by the filmmaker.

Through a study conducted on grain and speed, and while employing the interbellum's scientific cinema, Philippe Cote and David Dudouit pursue in the documentary format the plastic investigations initiated by structural cinema. In doing so, they take part in the continuity of, according to Nicole Brenez,[2] the "descriptive poems" of filmmakers such as Peter Hutton and Robert Fenz. They therefore expand the approach of experimental films dedicated to landscape since the 1960s which, according to P. Adams Sitney, associate "the representation of natural beauty with explorations of the way in which the film-making tools 'see' and 'know' their subjects".[3] The lyrical expression that is present throughout their films indeed highlights a patient observation of nature, revealing unique rhythms and forms through experiments that aim to enable a dialogue between the properties of the film and those of the landscape. This exploration of nature by the specific means of cinema links back to the first avant-garde's (Dulac, Epstein, Deluc) interest in the perceptual qualities of such medium, which was envisaged as a real instrument of knowledge by its ability to *make visible* beyond the visual limits of the human eye.[4] Éric Thouvenel detects, however, in the animistic and almost magical character that Epstein attributes to cinema, through notions such as the "photogenic", "a reaction to the positivism that so heavily weighed on the nineteenth century. [...] a desire to restore a part of mystery to a world burdened by rationalistic mortgage."[5] This Epsteinian critique of rationalism, of which Ken Slock has notably analyzed the relationships with the phenomenology of Maurice Merleau-Ponty,[6] does indeed present itself throughout the theoretical work of the filmmaker, who aims to reconcile reason and feeling, science and aesthetics.[7]

L'Angle du monde and *Île de Sein*, through an approach that is descriptive whilst poetic, beckon us to apprehend them in the light of the above-mentioned "re-constitution of *knowledge*"[8] by cinema, and to ask ourselves to what extent Cote and Dudouit propose, by soliciting the perceptual modalities of the filmic device, a sensitive and dynamic exploration of natural phenomena, thus asserting a departure from the categorical and linear approach on which science has developed. To this end, I will centre my analysis on the manifestation of cloud phenomena, to which the filmmakers grant a significant place and which are presented through their relationship to the sea and the land. Particularly abundant in this humid region and heavily exposed to ocean currents, they exist in a variety of shapes and movements, and, in what can be seen from the opening of both films, a mass of dense, and either stationary or mobile, clouds. By its lack of defined contours and its constant movements, the cloud also appears as a conducive figure to a reflective investigation of the properties of cinema as well as of the formation of images. Creating an indistinct space and working the instability of landmarks, it indeed confronts the viewer with the physical and mechanical nature of the filmic device.[9] Anouchka Vasak observes that the cloud, like the "meteor", leads to "the path of another mode of perception, diffuse and without border", inviting us to think in terms of the fluidity and the chaos that characterize the climate, and to "grasp phenomena, in the broad sense, within the horizon of their appearance"[10] in affinity with the phenomenological mode of perception.

By describing how the cloud and the film converse, affect each other and reveal themselves, I will analyze how the filmmakers, through their formal choices, manage to make the dynamic and microphysical processes of clouds perceptible, but also how these latter, by exacerbating and highlighting the properties of the image, manifest their presence and intensity. I will attempt to show how this dialogue between the physics of the film and the physics of the clouds testifies and participates in a global aesthetics that I would qualify as nebulous. Referring to what is "covered by clouds" and by analogy what is "veiled", "blurry", "imprecise",[11] this term will allow me to describe a poetic form which, based on openness, transformation and symbiosis, questions the boundary between film and reality, man and his environment and summons the part of uncertainty of a nature presented through its mobility and its incessant variations. I will articulate my analysis around three parameters linking the film to the cloud: time, space and matter.

The Time of Clouds

Shifting, changing, suspended, the cloud is a privileged object of frame-by-frame recordings, especially within scientific cinema. As a matter of fact, the time-lapse, which first took place in scientific experiments on movement, is a technique that manages to capture the dynamic processes of cloud physics.[12] By this process, *Île de Sein* as well as *L'Angle du monde* restore, in the words of Éric Thouvenel on the cinema of Jean Epstein, "the course of the clouds to our scale", giving us "access to a 'telluric' vision, a macroperception of events".[13] Effectively, Epstein's aesthetics were, during the first cinematic avant-garde movement, considerably influenced by scientific cinema.[14] Acceleration, Thouvenel further observes, is a process that in Epstein's world lightens and "rids [the cloud] of [its] little mass", now rendered "even more aerial".[15] In the films of Cote and Dudouit, the cloud also appears more evanescent and more fluid: "Yes, even when I film clouds… It can be a fluidity that is linked to the liquid element. The clouds become purely a flow…."[16] The temporal effects in these films are not, however, limited to providing details and movements invisible to the naked eye, and thus go beyond their strict scientific aim.

Recorded frame by frame over a long period of time, "often several hours, or even entire days",[17] the shots of *Île de Sein* seem to subject the entire landscape to this cloud temporality, making the movement of men, waves and other elements that compose it discontinuous. They seem to thereby subordinate the fate of the *Sénans* to the atmospheric movements that regularly expose this island to violent storms.[18] Whilst applied almost systematically in *Île de Sein*, the "play of temporal perspective"[19] is less predictable in *L'Angle du monde*, where not only time-lapses but also slow motions are applied in an irregular and intuitive way. For instance, in a sequence at the end of the film, whilst presenting a sunset over the ocean and cumulus clouds of various sizes, the filmmaker suddenly multiplies the abrupt and unpredictable changes of pace during the shot. The clouds thus pass quickly in front of the sun, before slowing down, resuming their course, freezing, etc. This "diversification of time",[20] unique to cinema, which the filmmaker solicits in the same shot and through a sensitive rapport with the camera mechanism, can be placed in relation to the variable image frequencies that characterize the first silent films, and which depended on the rotation speeds of cranks performed by the operator. This diversification, furthermore, asserts the unpredictability of clouds and therefore contradicts attempts to forecast the weather. Of varying sizes and densities, clouds are indeed unstable and complex systems, constantly interacting with the atmosphere (temperature, wind, humidity), which alters their particles and speed. Such challenge in terms of modelling presents itself as the main source of uncertainty when it comes to predicting global warming.[21]

The camera's capabilities to record temporally thus seem, according to both Dudouit and Cote, less to serve to collect data and empirical measurements than to highlight the overall movement of a landscape in perpetual becoming through a non-instrumental use of the mechanical means of film. As Nicole Brenez rightly notes: "From the visual initiatives of scientific experimentation, avant-garde cinema tears all filmic speeds from their machinic and functional determinations to reveal their formal virtues, to free their functions, to enliven their concrete sources".[22] Their application in films thus testifies to a deployment of the fundamental properties of the filmic device which, by returning to its own primitiveness, appears to thrive to make perceptible an elementary and primitive dimension of the landscape. This primitiveness is also invoked at the beginning of the analyzed sequence of *L'Angle du monde*, and which is marked by a luminous flickering of the image generated, according to Chris Kennedy, by "the combination of time-lapse and the camera's automatic exposure".[23] The brightness thus changes from one frame to the next, depending on the movement of the clouds in front of the sun, and the flickering stops when the latter is completely obscured by the cloud masses. Whilst highlighting the photogrammatic structure of the film, this flickering that runs through a significant part of the time-lapsed shots also refers to the mechanical system of the film projector, whose shutter work seems to be metaphorized by this passage of the clouds, between the filmmaker's gaze and the sun. Placed between the lantern and the film projection corridor, the shutter cuts the light beam during the transition from one image to another. It thus produces a beat that ensures an illusion of movement by masking the scrolling of the film and the interval between the frames. In the early days of cinematography, the shutter only blocked the light once per frame, causing a flickering effect on projection.[24] Additional shutter phases subsequently eliminated this phenomenon, which can still occur when the shutter is not adapted to the film's pace. In this respect, the sun that projects its light and the irregular passage of clouds seem to poetically represent an archaic or crisis-ridden projection device, no longer producing the illusion of continuous movement or homogeneous light. What appears in a relationship of equivalence between the movement of the shutter and that of the clouds, as a projection of the world onto the filmmaker—or, as one might say, a projection of the filmmaker by the world—indeed testifies, through these cadences and this variable flickering, of a great instability, as if Philippe Cote was expressing here his desire to reintroduce uncertainty within a predictable and regular mechanism. This game on speed, being deliberately open to the variations of nature, thus makes sensitive the various imperceptible changes that affect the landscape and the chaotic and non-linear dimension of the climate, notably described by the American meteorologist Edward Lorenz, who takes note of the great sensitivity of complex systems, an example of which manifests in clouds' sensitivity to the initial conditions at the time of their formation.[25]

An Open World

Aristotle observes in his *Meteorologica*[26] that clouds, like mirrors, are only apparent in what they reflect, which are light and colours. They also constitute the visible element of the water cycle's atmospheric phase and represent signs of phenomena such as wind or storm, thus attesting to an indicative value with respect to their environment. As seen in the films of our corpus, this connectivity is created or highlighted by the use of timelapse as well as black and white and contrasting colours, which put the sky and the Earth in dialogue, moulding the landmarks' instability and the landscape's transformation.

In *L'Angle du monde*, the clouds are in constant change and significantly alter the rest of their environment. If the clouds' exposure can vary from one shot to another, or, as it may even appear from one's perspective, between frames, their movement also has an impact on the brightness of the Earth and the sea. In a sequence filled with blue and grey colours, "fruits of the encounter between the impressed [colour] film and the sought low lights",[27] accelerated clouds of varying density cause a shift in the light's reflection on the water as they pass in front of the sun, whilst projecting their shadows onto the sea (Fig. 12.1).

In *Île de Sein*, the clouds are particularly related to the sun's motion of rising or setting, as it illuminates their forms in different ways. In some shots, they appear as a black strip similar to that formed by the Earth, separated by a white strip of sky. Here the distinction between the

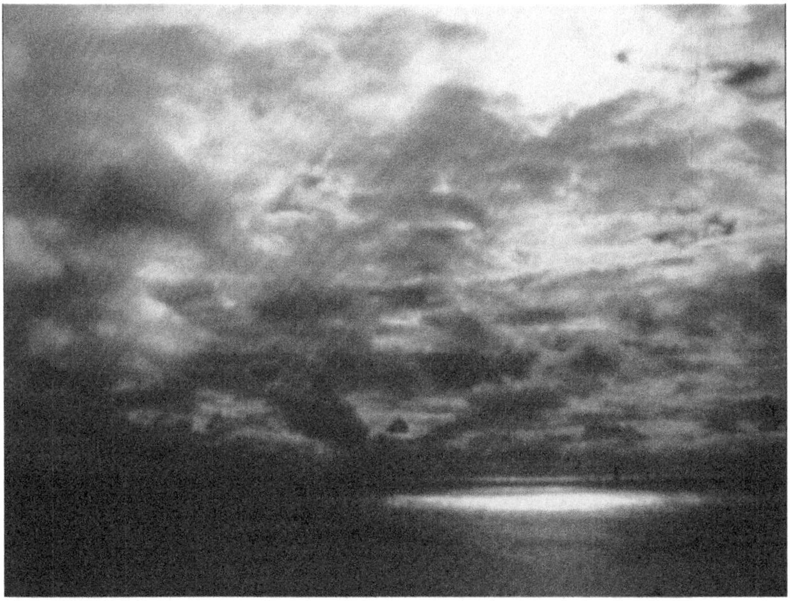

Fig. 12.1 Philippe Cote, *L'Angle du monde* (2006) © All rights reserved by the artist/Courtesy of Light Cone

upper and the lower is subverted, as the two spaces even merge themselves together in a striking shot where the clouds seem to touch the Earth, and form, with the effect of the diminishing light, luminous curves evoking the foam on the shore, before completely blending together and disappearing in an image that has become black (Fig. 12.2).

These interactions, through which the cloud is connected to, or even mixed with other elements, create an uncertain and nebulous space, accentuated by the images' underexposure and by the movement's discontinuity induced by the use of time-lapse, which prevents the viewer from locating themselves in the space of stable landmarks. The last example above, which concerns the convergence of the black masses that form the sky and the Earth whilst evoking the island's risks of flooding—linked notably to its low height and the rise of water due to global warming[28]—also helps assert a trend towards abstraction. By masking the figurative content of the landscape, this mass manages to render the overall movement perceptible. Regarding the filmmaker Henri

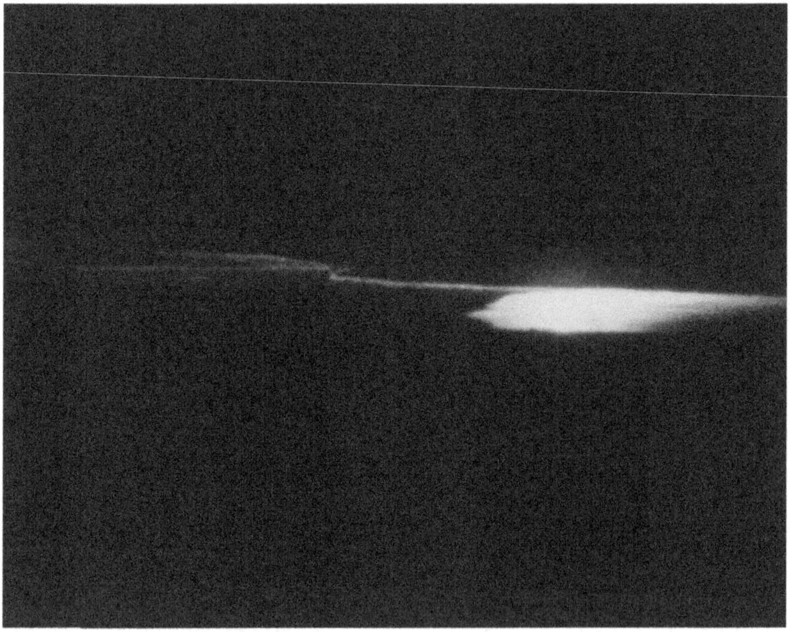

Fig. 12.2 David Dudouit, *Île de Sein* (2009) © All rights reserved by the artist/Courtesy of Light Cone

Chomette, Éric Thouvenel describes a similar approach: "it is not a question of moving away from reality but of retaining, through abstraction, only the trace, and almost the spectre, which precisely was not visible to the naked eye but whose existence cinema reveals".[29] The white line that separates the sky and the Earth can symbolize, in this sense, an opening, like a wound giving access to the new order of reality, to which, according to Jean Epstein, cinema introduces us,[30]at the same time that it presents the filmmaker's own openness to the world. This wound is also that of the image itself, whose abstraction, as it asserts itself, reminds us of its plastic qualities and its medium. But also of its grain, its movement and its rhythms, through which the filmmakers expand and figure their perceptual experience of the landscape. Jean-Michel Durafour, whilst trying to define the figural in cinema, speaks of a "lesion", an "injury"[31] made to the image by phenomena that manifest it "less as a representation than as a presence"[32] according to the terms of Philippe Dubois. A figure detached from the discursive order, the figural and the space it establishes define well this image which, whilst remaining attached to reality, becomes an "object of experience ... of the cinema-matter",[33] making "sensitive to the organicity of materials, to the fluidity of spaces, to the modulations of form and the formless, to the effects ... of what is neither sense nor resemblance".[34] The cloud, by giving, according to Hubert Damisch, "access to a world of shapes in movement forms and deformed by movement",[35] contributes to the deployment of this figural space, whilst also stimulating the viewer's interpretative and imaginative process, to which it provides an "incomparable material".[36] The cloud indeed establishes a continuous process of deformation of the image which, according to Bachelard, characterizes the very work of imagination:

> If there is no change, or unexpected fusion of images, there is no imagination; there is no *imaginative act*. ... [I]magination is essentially *open* and *evasive*. It is the human psyche's experience of *openness* and *novelty*.[37]

This imagination appears to reach its full deployment in the dark, on which the shot opens, possibly indicating its limits or its breakthrough outside the scope of visibility. The image here, by the plasticity of its forms and its gradual disappearance, appears as an invitation towards "extending the domain of 'that which is depictable'"[38] through the film, and also to open this latter to the immateriality of the cloud, to the spectator's imagination, to the filmmaker's interiority, to what is beyond or beneath any image. This attraction to the cloud emerges therefore as a penchant for the formless, the invisible and the impalpable, employing a reflection on the image of cinema as threshold between the apparent phenomenon and what remains hidden, between the subject and the world, man and his environment. By blurring the boundary between these different dimensions (through the absence of image or abstraction), the film manifests their symbiosis and returns them to the heart of the same matrix, whose figure is constituted by the island. A purified place, surrounded by water, exposed to telluric forces, the island indeed appears as an elementary, primitive space, a generic, immutable island, filmed by a camera that is often fixed and distanced, like a virgin eye casting a gaze "without prejudices, without morals, exempt from influences"[39] on the world. This original landscape, possessing a tendency towards abstraction, thus encompasses all at once the filmmaker, the spectator and the filmic device. If they find themselves in front of the ocean or on the sea, they are also immersed in the world, caught in this amniotic fluidity of clouds and cosmos that reveals them to themselves.[40]

Silver Crystals and Water Droplets

In various shots of *L'Angle du monde*, the view of the landscape is obstructed by water droplets placed on the camera lens. These images, which testify to a permeability, or even a certain vulnerability of the device whilst faced with reality and its hazards, relate to the very composition of clouds, which involve water particles or ice, also known as hydrometeors. One of these shots presents especially waves which, crashing against the rocks, rise and mingle with the clouds, projecting a thin rain on the camera. All whilst asserting the natural specificities of Brittany, "land of standing stones raised against the wind and the force of constantly active elements",[41] in the words of Sophie Gondolle, this shot evokes the phenomena that precede the formation of clouds, and more specifically the spray formed by the breaking waves whose saline nuclei—which act as catalysts for steam condensation—are driven by the ascending currents. These shots, furthermore, attest to an attraction to outburst and dispersion, already observed by Philippe Dubois in the cinema of Jean Epstein through the storm figure, which manifests, according to him, "a taste ... for everything that relates to outburst (of the bursting and the splashes) of matter, and a certain fascination for the destabilizing effects of (fixed) forms".[42] In both *L'Angle du monde* and *Île de Sein*, this attraction is also emphasized in the palpable swarming of the image's grain, generated by the random arrangement of halide crystals on the film emulsion. Exacerbated by the film's sensitive reactions to light, this swarming is added to the continuous movement of ice crystals and water particles, whose state and position constantly change. These elements condense, merge, evaporate, freeze, melt or rub under the influence of atmospheric conditions, generating lightning or precipitation in the form of rain, snow or hail.[43] In this sense, this visible agitation of the grains reveals an analogy between the cloud and the film, whose movement thus affirms the dynamic character of the hydrometeors. In a text dedicated to the series of *Equivalents* (1925–1934) by Alfred Stieglitz, Philippe Dubois also notes the equivalence between the indicative nature of the cloud and "that of the photographic sign".[44]

The repetition of the cloud and the water particles by the silver crystals is particularly evident in a shot from *L'Angle du monde* where the film-maker, heading towards Molène and Ouessant, observes from the deck of the boat as the island of Sein recedes and disappears behind a veil of water formed by sea spray. The image grain, very pronounced, then looks like it is becoming part of the sky and the fine rain that has now pervaded the frame, as the still distinguishable figures (the coast, the light-house) seem threatened to be dissolved within the nebula formed by silver salts. This shot, Philippe Cote also explains, "demonstrates dialectically opposing qualities: on the one hand, the stable (the lighthouse, the land, the boat), and on the other hand, the unstable and the fluid (the swell, the spray)".[45] It is interesting to observe that this play of opposites also characterizes the phenomena of warming and cooling that preside over the process of cloud formation, particularly those that take place by the sea. As explained by Benoît Vie, "when a sea breeze comes to counter the dominant [warmer] wind, this forces an uprise at the convergence zone and can lead to the formation of ... cumulus".[46] Such friction pertains to a dynamic that characterizes his films' relationship with reality, and which, by the aggravation of the grains and the explicitation of the processes related to the movement of still images, brings together the stable and the fluid, the inert and the living, the crystal and the cloud.

The nebulous aesthetics that characterizes *Île de Sein* and *L'Angle du monde* therefore manifests itself in a form of indistinction, which forges a confusion between the film and a nature that comprises infinite variations. Presented in a relative space–time, a blurry image and a matter that is exacerbated by the film roll's reactions, the landscape is subject to multiple metamorphoses, brought to the boundary of abstraction by a device equipped with its fragility and its fundamental components. In a text dedicated to Jean Epstein and avant-garde cinema, Nicole Brenez speaks of a "strangeness" of the phenomena that cinema reveals. She writes: "this strangeness does not indicate a mystery, or something unmentionable and obscure ... but an essential alteration, the deeply unidentifiable and impure dimension of phenomena, which cinema detects, welcomes and elaborates ...", whilst keeping up with the "unstable, disordered, relative and unintelligible nature"[47] of things. What these films depict, in their dialogue with the clouds that in turn affect them and reflect them like a mirror, is perhaps, above all, this strange "alteration", this tipping of things towards another state, this capsizing of the image towards the elements, and their reciprocal metamorphosis. These different tipping points are tied to the atmospheric and climatic variations that affect and transform all living things, questioning the common outcome of man and the world. Through the medium of film but also through the unstable and mobile landscape, Philippe Cote and David Dudouit summon our intimate, sensory and primal relationship with the depicted elements, placing us in a state that is not simply about paying attention to these, or about allowing us to be better acquainted with, apprehend or even dominate them, but rather one that involves connection and relinquishment in the face of an elusive nature.

Notes

1. L'Abominable is an artist-run film laboratory, located in La Courneuve, which provides artists with tools related to film. See their website: https://www.l-abominable.org/.
2. In the words of Nicole Brenez in "Flux, flash et prisme: Vitesses critiques dans l'histoire du cinéma", *Le Genre humain*, no. 49, Paris, Éditions du Seuil, 2010, pp. 125–138. Available online at: https://www.cairn.info/revue-le-genre-humain-2010-1-page-125.htm [consulted on 26/01/2022]. Unless otherwise noted, all translations from original French by the author and translators.
3. Paul Adams Sitney, "Landscape in the Cinema: The Rhythms of the World and the Camera", *in* Salim Kemal and Ivan Gaskell (eds), *Landscape, Natural Beauty and the Arts*, Cambridge, Cambridge University Press, 1993, p. 125.
4. See on this subject: Viva Paci, "Chapitre 2. La nature du cinéma: entre photogénie et attraction", in Viva Paci, *La machine à voir: À propos de cinéma, attraction, exhibition*, Villeneuve d'Ascq, Presses universitaires du Septentrion, 2012. Available online at: https://books.openedition.org/septentrion/9463?lang=fr [consulted on 31/08/2021].
5. Éric Thouvenel, *Les images de l'eau dans le cinéma français des années 20*, Rennes, Presses universitaires de Rennes, 2010, p. 189.
6. Ken Slock, *Corps et machine - Cinéma et philosophie chez Jean Epstein et Maurice Merleau-Ponty*, Sesto San Giovanni, Mimesis, coll. "Images, mediums", 2016.
7. "Against the aegis of reason, he will therefore speak, with fervour, of the feelings that cinema awakens, implacable, overwhelming. Through him, the revealed truth asserts itself more tenaciously,

more intuitively than the truths constructed and balanced under the regime of the rational", in K. Slock, *ibid.*, p. 81.

8. *Ibid.*, p. 9.
9. In *Clouds* (1969) by Peter Gidal, the cloud serves for example as a support for the filmmaker's anti-illusionist project.
10. Anouchka Vasak, "Penser météore", *Communications*, Vol. 2, no. 101, 2017, pp. 7–20. Available online at: https://www.cairn.info/revue-communications-2017-2-page-7.htm [accessed 31 August 2021].
11. CNRTL, "Nebuleux", online at: https://www.cnrtl.fr/definition/n%C3%A9buleux [accessed 31 August 2021].
12. See for example the film by Japanese meteorologist Abe Masanao: *Fujisan komo no ugoki* (*Mount Fuji—The Movement of Clouds*), Japan, 1929.
13. Éric Thouvenel, *Les images de l'eau dans le cinéma français des années 20*, *op. cit.*, p. 204.
14. On this subject, see: Maria Ida Bernabei, "L'accéléré dans la théorie des années 1920: 'Germination du blé… Cristallisation… Films documentaires et, surtout, films d'avant-garde'",*1895*, no. 87, 2019, pp. 90–111. Available online at: https://www.cairn.info/revue-1895-2019-1-page-90.html [accessed 29 January 2022].
15. *Ibid.*
16. Philippe Cote, " À la recherche de l'image cachée", interview conducted by Michèle Blumental, 2017. Available online at: http://derives.tv/a-la-recherche-de-limage-cachee/ [accessed 31 August 2021].
17. See the introductory text about David Dudouit written by Stefano Canapa and Guillaume Mazloum at this URL: https://lightcone.org/fr/cineaste-3773-david-dudouit [accessed on 31 August 2021].
18. Annaig Oiry notably explains on this subject: "The Finistère island, extending the Pointe du Raz off the southwest of Brittany, is in many respects an extreme case … for the study of natural risks on the metropolitan territory: a low island (it only emerges at an average height of 1.5 meters, with a highest point at 9 meters), small (about 60 hectares), strongly indented and exposed to the 360 degrees of the compass rose, it is particularly vulnerable to storms", in the article "Vivre avec ou par le risque à l'île de Sein

?", *EchoGéo*, no.19, 2012. Available online at: https://journals.openedition.org/echogeo/12960 [accessed on 31 August 2021].

19. According to the expression of Jean Epstein, in *The Intelligence of a Machine* [1946], trans. Christophe Wall-Romana, Minneapolis, Univocal Publishing, 2014, p. 18.
20. Jean Epstein, "The Fluid World of the Screen" [1950], trans. Sarah Keller, in Sarah Keller, Jason N. Paul (eds), *Jean Epstein. Critical Essays and New Translations*, Amsterdam, Amsterdam University Press, 2012, p. 385.
21. Clouds, indeed, play a central role in the climate. As explained by Camille Risi and Sandrine Bony: "On the one hand, they cool the Earth, by shading solar radiation. On the other hand, high clouds warm it, by contributing to the greenhouse effect." However, the two researchers remind us, clouds, due to their variable sizes, do not all fit into the climate models used which "represent the circulation of the atmosphere on grids that measure about 100 kilometres". Thus "we can only represent the average effect on each grid of a set of clouds" and depending on the simplifying assumptions used, the climate models "lead to different results", in Camille Risi and Sandrine Bony, "Les nuages, enfants terribles du climat", *The Conversation*, 2019. Available online at: https://theconversation.com/les-nuages-enfants-terribles-du-climat-113102 [accessed on 31 August 2021].
22. Nicole Brenez, "Flux, flash et prisme", *art. cit.*
23. Chris Kennedy, "Philippe Cote, Reflection", 2018. Available online at: http://derives.tv/philippe-cote-reflection/ [accessed 31 August 2021].
24. This flickering is also named by Fleur Chevalier "Projection flicker" in "Cinéma Argentique et 'Flicker': Le cas particulier des films dits d'avant-garde ou expérimentaux", *ARTOBS*, Available online at: https://obsolescence.hypotheses.org/files/2011/03/Flicker_Fleur_Def2.pdf [accessed 31 August 2021].
25. "A complex system is defined as a system composed of a multitude of entities whose local interactions give rise to global properties that are difficult to predict from the sole knowledge of the properties of these entities. A flock of birds, a social network, ecosystems, individuals in motion or even a network of neurons are complex systems", CNRS, "Qu 'est ce qu'un système complexe ?", *L'Incubateur de Médiation Scientifique du CNRS.*

Available online at: https://iscpif.fr/incubateur_mediation_cnrs/portfolio-item/quest-ce-quun-systeme-complexe/ [accessed 26/01/2022]. See also on this subject: Pierre Bergé, Monique Dubois, "CHAOS, physique", *Encyclopædia Universalis*. Available online at: https://www.universalis.fr/encyclopedie/chaos-physique/ [accessed 31 August 2021].

26. See Chapter IV of Book III of Aristotle, *Meteorologica*, trans. H. D. P. Lee, Cambridge, Mass., Harvard University Press, 1951.
27. Philippe Cote, "*L'Angle du monde*: Fabrication d'un film tourné à Ouessant, Molène et Sein", *in* Roxane Hamery and Éric Thouvenel (eds), *Jean Epstein: Actualité et postérités*, Rennes, Presses universitaires de Rennes, coll. "Le Spectaculaire", 2016, p. 303–308. Available online at: https://philippecote.wordpress.com/category/mes-films-photogrammes/ [accessed 31 August 2021].
28. According to the oceanographer Paul Tréguer: "the island of Sein is threatened with submersion by the end of the century if we continue the current warming regime". Paul Tréguer, "L'île de Sein est menacée de submersion", *Le Télégramme*, 2015. Available online at: https://www.letelegramme.fr/bretagne/paul-treguer-l-ile-de-sein-est-menacee-de-submersion-24-10-2015-10824050.php [accessed 31 August 2021].
29. Éric Thouvenel, *Les images de l'eau dans le cinéma français des années 20*, *op. cit.*, p. 117.
30. "The cinematograph … shows time to be merely a perspective resulting from the succession of phenomena, the way space is merely a perspective of the coexistence of things. … Thus, after having taught us about the unreality of both continuity and discontinuity, the cinematograph rather abruptly ushers into the unreality of space–time", Jean Epstein, *The Intelligence of a Machine, op. cit.*, p. 21–22.
31. Jean-Michel Durafour, "Le Cinéma", *in Jean-François Lyotard: questions au cinéma. Ce que le cinéma se figure*, Paris, Presses Universitaires de France, coll. "Intervention philosophique", 2009. Available online at: https://www.cairn.info/jean-francois-lyotard-questions-au-cinema--9782130576983-page-69.htm [accessed 31 August 2021].
32. Philippe Dubois, "La question du Figural", in Pierre Taminiaux and Claude Murcia (eds), *Cinema, art(s) plastique(s)*, Paris, L'Harmattan, 2003, p. 54, quoted by Jean-Michel Durafour, *ibid.*

33. Philippe Dubois, “La Tempête et la matière-temps, ou le sublime et le figural dans l’œuvre de Jean Epstein”, in Jacques Aumont (ed.), *Jean Epstein, cinéaste, poète, philosophe, conférences du Collège d’histoire de l’art cinématographique*, Paris, Cinémathèque française, 1998, p. 270.
34. Philippe Dubois, “L’écriture figurale dans le cinéma muet des années 20”, in François Aubral and Dominique Chateau (eds), *Figure, figural*, Paris, L’Harmattan, coll. “Ouverture philosophique”, 1999, p. 248.
35. Hubert Damisch, *Theory of /Cloud/. Toward a History of Painting* [1972], trans. Janet Lloyd, Stanford, Stanford University Press, 2002, p. 19.
36. *Ibid.*
37. Gaston Bachelard, *Air and Dreams: An Essay on the Imagination of Movement* [1943], trans. Edith R. Farrell, Dallas, The Dallas Institute Publications, 1988, p. 1.
38. Hubert Damisch, *Theory of /Cloud/, op. cit.*, p. 90. The author borrows the concept of “depictability” [*Darstellbarkeit*] from Sigmund Freud, *The Interpretation of Dreams* [1900], trans. A. A. Brill, Hertfordshire, Wordsworth Editions, 1997.
39. Jean Epstein, “The Cinematograph Seen from Etna”, trans. Stuart Liebman, in Sarah Keller, Jason N. Paul (eds), *Jean Epstein. Critical Essays and New Translations, op. cit.*, p. 292.
40. For Emanuele Coccia, climate also seems to designate this matrix: “The climate is not the collection of the gases that envelop the terrestrial globe. It is the essence of cosmic fluidity, the deepest face of our world, the one that reveals it as the infinite mixture of all things, present, past, and future. The climate is the name and the metaphysical structure of mixture. … … In all climates, the relation between the container and the contained is constantly reversible: what is place becomes content, what is content becomes place. The medium becomes subject and the subject becomes medium. All climate presupposes this constant topological inversion, this oscillation that undoes the border between subject and environment, a role-reversing oscillation”, in Emanuele Coccia, *The Life of Plants: A Metaphysics of Mixture* [2016], trans. Dylan J. Montanari, Cambridge, Polity Press, 2019, p. 27. Quoted by A. Vasak, “Penser météore” *art. cit.*

41. Sophie Gondolle, "Jean Epstein et la Bretagne, un ancrage légendaire" in R. Hamery and E. Thouvenel, *Jean Epstein*, *op. cit.*, p. 202.
42. Philippe Dubois, "La Tempête et la matière-temps, ou le sublime et le figural dans l'oeuvre de Jean Epstein", *art. cit.*, p. 269.
43. See on this subject, Benoît Vie, "Que se passe-t-il dans les nuages ?", *Encyclopédie de l'Environnement*, 2021. Available online at: https://www.encyclopedie-environnement.org/air/physique-nuages/ [consulted on 31 August 2021].
44. Philippe Dubois, *L'Acte photographique et autres essais*, Paris, Nathan, coll. "fac. image", 1990, pp. 189–190.
45. Philippe Cote, "*L'Angle du monde*", *art. cit.*
46. Benoît Vie, " Que se passe-t-il dans les nuages ?", *art. cit.*
47. N. Brenez, "Ultra moderne. Jean Epstein contre l'avant garde (repérage sur les valeurs figuratives)", in Jacques Aumont (ed.), *Jean Epstein, cinéaste, poète, philosophe, op. cit.*, 1998, p. 216.

CHAPTER 13

Enhanced Detection Through the Stochastic Properties of Terrestrial Fluids

Bidhan Jacobs

Two complementary conceptions of film technologies, related to perceptual and cognitive sensitivity, can be derived from Jean Epstein and Jacques Perconte. Epstein believes that human capabilities need to be enhanced by the instruments of cinema.[1] Since the sensory apparatus only operates on a tiny bandwidth below and beyond which phenomena are neither visible nor audible (too high or too low frequencies), nor analysable (excessive or insufficient amount of data), it needs to be supplemented by the film machine to extend its knowledge of the infinity of possibilities. This is why Epstein thought that "the camera lens is ... an eye endowed with inhuman analytical properties",[2] the only one capable of drilling into the real. On the contrary, Perconte warns us that the most advanced of technologies, programmed and programming, will always be reductive, and its computing power will always be surpassed by the perceptual powers of the human being: "Digital technology has given our

B. Jacobs (✉)
National School Supérieure Louis-Lumière, Saint-Denis, France
e-mail: bidhan.jacobs@gmail.com

E. Della Noce and L. Murari (eds.), *Expanded Nature*, Experimental Film and Artists' Moving Image,
https://doi.org/10.1007/978-3-031-70729-2_13

machines the means to reduce the world, which is essentially unquantifiable, into a given quantity of information. Our images can only be systematic, partial, imperfect …. It's life in low definition, always."[3] And to conclude: "technology will not be able to see what I see".[4]

One solution would be to rethink recording systems (detection) and enhance them for the purposes of formal inventions aimed at giving the viewer more accurate perceptual experiences of complex physical and psychic phenomena.

Starting with the seminal film by Lumière operator Félix Mesguich in 1900, *Nice: panorama sur la ligne de Beaulieu à Monaco, (Film Lumière n°1230 et 1932)*[5]—though preceded by Eadweard Muybridge's magnificent photographs of cloudy skies from the late 1860s and by Étienne-Jules Marey's aerodynamic studies of smoke currents in 1899—one of the strategies employed by certain filmmakers and artists is to immerse their detection tool within the elements of nature, in a gaseous or liquid state—terrestrial fluids—so that light, before its optical transformation within the camera lens, undergoes immensely complex deviation phenomena governed by fluid mechanics, thermodynamics and so many overdeterminations that, for example, the powerful probabilistic mathematical models developed for meteorology or the oceans still do not resolve them.

We can trace, from Jean Epstein—and his work on the Breton islands—a history of these films that have integrated the phenomena of the dynamic variation of fluids and their stochastic properties.

The foundational film for the ongoing experimentation on the stochastic properties of fluids was *Finis Terrae* (1929) by Jean Epstein, which marked the filmmaker's break with the film industry within which, at the cutting edge of the avant-garde, he had achieved recognition with his masterpiece *The Fall of the House of Usher* (1928). Epstein broke away from Paris for Brittany, leaving the studios for a marine cinematographic laboratory in order to renew processes and forms. As he put it: "I set off to film the men of the four elements, the kelp-burners; those who dip their arms in the sea and fire, who sail on the breezes, and stretch out to sleep on the slightest exposed land. If there is an alchemy, the soul of such men cannot disappoint. It does not disappoint."[6] Epstein retains these four elements: fire—smoke, solid particles that also connote the earth; water—the sea, its vaporous forms of foam and mist; air—through the wind that pushes and stirs the opaque or translucent particles in suspension, whose fluid qualities he observes interacting according to their viscosity, creating constantly renewed conditions of volatile optical properties. The opening and conclusion of the film are remarkable in this respect and should be seen in superimposition.

The opening sequences show the three sailors on the deserted island of Bannec burning whole rows of the seaweed that they harvest regularly to transform it into precious caustic soda. Epstein conducts three main studies: the first is the effect of light on these thick, large volutes that twist to resemble clouds, whose density changes with the wind, modifying in parallel the density of the dazzling and opaque white, until it becomes weak and evanescent. The second study focuses on the bodies of the three sailors, made malleable: they are reshaped or become vaporous; their fragility in the work is confirmed. The third study focuses on the sun itself, detected through more or less dense and thick curtains of smoke stretched and carried away by the wind. The solar disc pulsates, the result both of the variable diffusion imposed by the smoke and of the semi-obscuration effect of the clouds of particles.

In the last section, the village women of Ouessant worry for the three men of whom they have no news. One is indeed injured and at risk of gangrene. The intensity of their anxiety is conveyed by slow-motion waves that crash against the reefs in huge slow explosions of white foam, which Epstein studies through their saturation of the frame, opacity and disintegration into fine, shimmering water particles that fall as rain or dissipate as steam. From the two islands, men embark. The mist rises, sometimes naturally, often artificially recreated by the unconventional use of a diffuser screen and a multiplicity of filters of all shapes and sizes. These artisanal methods are supported by the deliberate use of loss of focus. Epstein studies the same effects of heterogeneous, light or very strong diffusion and recomposition on the landscape as on the men. Epstein therefore places the effects of smoke (fine solid particles of carbon) and those of water, in the form of waves exploding on the reefs or suspended (or simulated) particles, in the same aesthetic category. Not only does Epstein organize the effects of the unlimited optical phenomena transmitted by his lens according to a precise and unprecedented transcription, but he also establishes a comparative aesthetic analysis that provokes a vast upheaval of what we thought we knew about them. The four elements are thus united in an alchemical vision of the world.

The optical and viscous properties of smoke can be detected in the studio, by isolating it from other fluids (except air), without any turbulence or resistance from a body, in order to explore to the maximum its intrinsic specificities of complex expansion and propagation in dynamic space. It was according to this principle that Robert Cahen[7] created the installation *Traverses* in 2002 with the support of Le Fresnoy and the CICV Pierre Schaeffer. *Traverses* has been part of the collection of the Museum of Modern and Contemporary Art in Strasbourg since 2003 and of the Harris Museum in Great Britain since 2009. This installation takes the form of a digital colour projection, on a 4 m by 3 m screen suspended in the middle of a dark room, of a vertically framed image, like a painting, within which the opaque white slowly and partially allows traces of isolated human figures to emerge, only to then engulf them in cycles.

Traverses is the direct consequence of a previous work by Cahen, *Suaire* (*Shroud*, 1997) which conducted a comparative analysis of the history of "non-man-made" images and those of film and video, using both electronic and physical means, in a double video projection on a single cotton veil, of slow-motion faces in very slow superimpositions. But it is more like a reinterpretation, enlargement and amplification of the short sequence at the beginning of *Finis Terrae*. Thus, the iconography associated with the negative imprint of the Shroud of Turin, which represents for Cahen "the strongest image of man that [his] cultural memory has retained ..., a non-photographic image printed on fabric by a hand that we cannot recognise",[8] is diverted to reshape the body using the optical properties of smoke. The process consisted of filming against a blue background, in high light and with maximum overexposure, first people walking facing the camera, then, separately, curtains of white smoke. The radicality and novelty of the plastic forms come from Cahen's choice not to use the viscosity of the smoke directly on the actors' bodies, but to provoke their visual interaction in post-production in order to study as precisely as possible how the dense plastic forms of smoke could render the human figure labile.

Patrick Zanoli, then head of the CICV's Virtual Images Department, spatialized the various curtains of smoke using the powerful professional compositing and special effects software Inferno to create very sophisticated 3D overlays. This technique made it possible to obtain the very slow emergence of imperceptible traces that gradually transform into barely discernible unstable nebulae, then into evanescent fragments of vaporous human figures, whose auric power evokes *The Holy Face* (1658) by the painter Zurbarán. This *Holy Face* presents a nebula in the form of a diffuse ellipse within a suspended veil, raised in the centre by a pin and painted in trompe-l'oeil. It shows both the trace and the distant origin of the nebula.

Zurbarán chose the form of the sketch, imitating an imprint, to represent the face of God. The power of the painting lies in its formal economy: the material trace refers to the divine. Cahen, unlike the Spanish painter, does not work on transcendence but rather brings an image back to earth, a moving, marvellous image of man as he has never seen himself before.

In his project on the Madeira archipelago (started in 2012 and completed in 2014), Jacques Perconte simultaneously reconnects with Epstein's Breton islands and the Normandy coast of the Impressionists. *Chuva* (2012) is directly linked by the artist to *Impressions* (2012), in which he wanted to film large waves, strong winds on a raging ocean: all images that he had in mind and was never able to obtain,[9] except through figurative work. Moreover, this absence has not prevented other projects from being born; the film also invokes *Finis Terrae* in its way of placing the detection tool within the marine optical events caused by the ocean water's uprising through condensation into mist and its precipitation from the sky as rain, which transform solids and unify all visible into a single fluid.

As a prerequisite for each of his films, Perconte's perceptual sensibility transforms the natural phenomenon into aesthetic feelings and emotions, the technological detection having to subordinate itself to his impressions. As he puts it: "The filming is really the most important moment for me. That's when the films are made. I know when I shoot that I'm recording something that could become extraordinary." On the other hand, transferring to Madeira the culture he forged during his first long shooting experience dedicated to a landscape during his "Haute-Normandie" series (2011–2012), he knew that filming a landscape requires letting go of preconceived ideas and "concentrating on what nature offers [him] and never considering it as a spectacle .... It's the concentration, the intensity of the camera's presence that allows the images to reflect what the landscape offers to [his] senses."[10] Nature's and the universe's gift during those few minutes after Perconte's arrival near Funchal was a manifestation of the power of the oceanic and atmospheric fluids working the sun's light waves. Perconte immerses his detection tool and "focuses" it in these optical epiphanies. The tool and the lens in question are the same as for *Impressions*: a hybrid digital camera, the Canon EOS 7D dating from 2009, mounted with an EF 70-200 mm L series zoom lens,[11] products from the most advanced optical, mechanical, electronic and digital technologies of the time, of the highest precision, allowing him to film in full HD (1080p). These technical details are crucial, as they determine the very powers of detection: the signal transformation in the lens is optimal, with the behaviour of the optics in MTF[12] being very high across the entire frequency range, even in the very high spatial frequencies, thus achieving one of the highest resolution powers for professional photographic zooms in the early 2010s, allowing a very high average contrast ratio. Fluorite and UD lens[13] technologies, combined with fifteen groups of special glass lenses and anti-reflection and resistance coatings, ensure that the signal is transformed with a minimum threshold of "aberrations" and "losses", and that image points are obtained almost perfectly on the plane of the large CMOS sensor, throughout the focal length range of the lens. The electronics ensure real-time data communication between the lens's microcomputers and the super-microcomputer and high-performance firmware in the camera body; the software should be seen as an extension of the optical transformation procedure in the sense that it supplements its physical limitations and makes it "perfect". Given the monumentality of the optical micro-phenomena in the lens field, the performances of the equipment used by Perconte take on their full meaning.

The static shot, unusual for Perconte until 2011 and inaugurated in his "Haute-Normandie" series, but taken to the extreme in his "Madeira" series, is associated with the use of the long focal length of the zoom lens; thus he can drill down into a very narrow field within which he juxtaposes and amplifies, through the extreme depth of field obtained by diaphragming to the maximum, all the variations and interferences of reflection, refraction, diffraction, diffusion, absorption of the light signal by the fine raindrops and mist, in the axis of the digital camera, from its entrance pupil to the horizon. While the powers of the optical system are largely exceeded by the extreme and random concentration of fine spheres of water falling at too great a speed, not to mention the infinite number of those in suspension progressively filtering the field, rendering it incapable of separating each micro-phenomenon while millions occur between each quantification of the signal, its calculated properties allow the impeccable transmission of variations in light interactions. Any other less efficient optical system would have thus transmitted noise. The first three minutes of *Chuva* unfold, within the limits of technology, all the tiny gradations that lead, for example, the frail and tiny sailboat at the bottom of the frame to slowly vaporize, the island on the horizon to merge with the water, the ocean to join the sky. The sublime fluid forms of this blur resulting from simple transformation are confirmed by the end of the film, after Perconte's temporal compression processes have performed their work of revelation, where shimmering blue images of underwater visions then spread out.

Leighton Pierce[14] literally immerses his camera underwater, enclosed in a translucent hermetic case, using a method inherited from some experiments by Jean Painlevé (collaborator of Jean Vigo and Man Ray), such as his periscope shots, the studio aquarium and his underwater films *Sous-l'Eau* (1935). Leighton Pierce has been constantly discovering new forms of blur since the 1980s, working in the direct lineage of Epstein on the phenomenon of fluidity. Here are just a few films with evocative titles: *Memories of Water #21, 6, 27* (1997), *Water Seeking Its Level (St Pons)* (2002), *Evaporation* (2002), *My Person in the Water* (2006) or the installation *Convection* (2007). *My Person in the Water* is entirely underwater, except for the opening—a fleeting submersion that takes us through the dioptre and gives us the sensation of the calm surface of a lake from which Pierce's head emerges—and the end, when the tilted mini-DV camera rises for a moment to reveal the line of the dioptre that separates the water from the air, only to dive straight back into the depths. This film is about two bodies: that of a woman, who undresses in the water, and, more surreptitiously, that of Pierce, which the optical deviations treat differently. The camera first whirls under the surface, seen from an acute angle that transforms it into stretches of dark green shades in an increasingly lighter gradation, as the depth of the sheet of water opacified by fine particles is revealed, within which the light is delicately diffused. The female body is affected by this slight diffusion, traversed by some diffraction reflections, but never subjected to optical distortions. On the other hand, the materiality of the water, its density and viscosity, slows down her graceful movements and make her appear weightless, an impression amplified by the slowly rotating camera, while the effervescence of shiny air bubbles invades her as she swims. During the low-angle shots, her body is reflected on the surface of the water like in an undulating mirror that extends all around her in radiant dark shapes with the unstable contours of light and shadows projected onto the lake. Meanwhile, in superimposed shots, Pierce, who is observing his loved one, is instead shown from under the surface of the water, his head and torso recomposed by the very strong refractions and diffractions into distended, liquified forms. The film then reaches a peak of plastic refinement when it is no longer possible to distinguish the two people in the couple, who seem not only to merge but to become one with the water: refractions

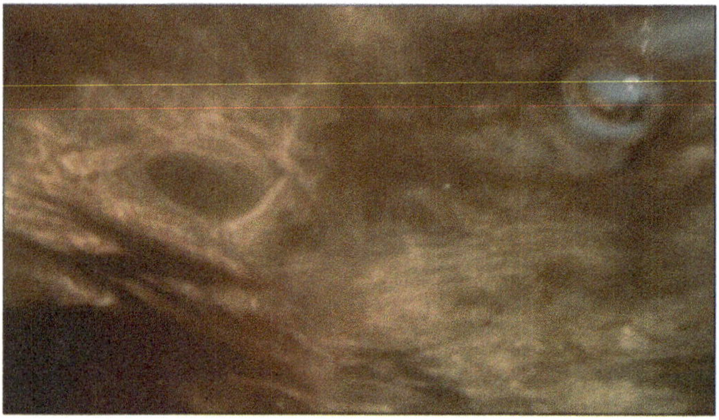

Fig. 13.1 Leighton Pierce, *My Person in the Water* (2006) © All rights reserved by the artist/Courtesy of Light Cone

and reflections of the bodies by the surface create these moving, entirely fluid human textures (Fig. 13.1).

In the end, all that remains is the water and, of the lovers, this shimmering volute rising from the depths. The beauty of the resultant fluid plastic forms—it should be noted that the low shutter speed, the transparencies and the loops amplify the optical phenomena detected—lies in the fact that they represent the infinite feelings of love.

Chott el-Djerid (A Portrait in Light and Heat) (1979) by Bill Viola synthesizes this history of adjoining the stochastic properties of fluids. *Chott el-Djerid* is part of a period when Viola's works were still produced by television. However, since joining the TV Lab, he has gradually moved from an approach based on technology as material (for which he was obliged to use television structures), to the understanding that he needed to include specific work on the system of human perception. He came to "consider video as a living system"[15] and then broke with signal processing to start working with real situations. He "first got rid of all the superfluous elements, to try and get back to what is fundamental. [He] ended up focusing [his] research, for several months, on the basic unit of video: the camera and the monitor."[16] Thus *Chott el-Djerid* is a pure work of detection, and one of the most striking in this history we are retracing of the means of exploiting the optical properties of fluids, related in this film to the convective movements of the scorching desert air in Tunisia. The video is motivated by the following question: "how can we get the camera to look inside our mind at what we cannot see with our eyes?".[17]

The work manipulates the only optical phenomena detected with very long focal length telephoto lenses (in particular an 800 millimetre lens for the desert) mounted on an electronic camera fixed to a tripod. After an opening studying the intense effects of the light on the snow-covered prairie in Saskatchewan, Canada, Viola, with the collaboration of Kira Perov, immerses his detection tool in the Sahara Desert at Chott el-Djerid, a lake that has become a salt flat (*chott* means shore in Arabic). Several short shots of the desert are cut to a wide shot of Viola throwing a stone into a water hole. Six shots follow to frame the water undulating under the stone's impact, then the red water with its surface speckled with whitish spots. During this sequence, the microphone records the wind's irregular blowing, the crunching footsteps, the plunge of the stone, followed by the ripples. A long superimposition transforms the matter of the moving water into a landscape of undulating coloured shapes. This crossfade is essential: it underlines an equivalence between the mechanical waves of water, those of sound and the detected electromagnetic waves. For Viola, reality is primarily wave-like (the matter constituting the crunching of steps—the grains of salt—just as the thrown stone reminds us of its corpuscular nature). Considering the lens as a detection tool is therefore particularly well suited to this work. The very fine field of its very long focal length lens drills several tens of kilometres of air near the ground (at a height of about one metre). Two main optical phenomena are detected in the camera's axis: first, the uneven reverberation of the ground which radiates the heat of the sun at zenith and animates the air into an infinity of rising volutes, a phenomenon that occurs from the surface of the lens to the horizon and twists the light rays into increasingly complex shapes as they travel several kilometres of convecting air; and second, the temperature differences of the air layers at altitude whose refractive indices become volatile, creating curvatures of the sun's rays and therefore the optical phenomenon of mirage. These two extremely disruptive optical phenomena interfere with each other. In addition, a third natural phenomenon is at work: larger-scale convections of the atmosphere in the form of wind, which causes air masses that are already convecting to move randomly, disturbing the temperature of the air layers and, consequently, the mirages. The viewer is never sure whether they recognize a human figure, a tree or a well, and is very

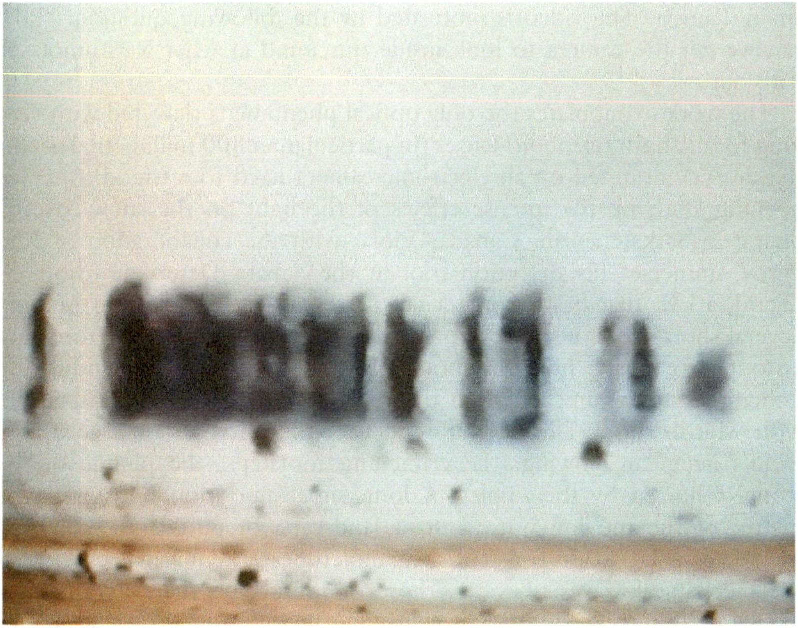

Fig. 13.2 Bill Viola, *Chott el-Djerid (A Portrait in Light and Heat)* (1979) Photo: Kira Perov © Bill Viola Studio

often deceived. From the horizon, small forms emerge that we assimilate to living beings. The first hypothesis—human beings; as they get closer to the camera, a second hypothesis is formulated, as we see legs taking shape—human beings riding animals; something starts to shine within this unknown shape as it moves and transforms—it's a reflection on a material that seems to be a pane of glass. Finally, an unstable figure of two motorcyclists riding one behind the other takes shape. Moreover, the mirages reverse spatial perception: we can no longer distinguish between the ground and the sky, everything is doubled in a horizontal symmetry. As aerial spectres advance, trembling and fragile, melting into an irregular sticky mass, their inverse double is also there, slender and oversized. The three natural phenomena bring forth a range of colourful undulating, wavering, pulsating, trembling, shivering, porous forms, organized in superimposed layers of turbulent fluids (Fig. 13.2).

The telephoto lens accurately transmits (albeit with some absorption as it was over 80 cm long) all the optical properties of the atmosphere in these conditions of extreme sunlight and heat. The impression of immersion in liquids is rigorously explained by physics. The strength of the video lies in this understanding of the matter of dynamic fluids that is beyond our senses. This meticulous description allows us access to simultaneously microscopic and macroscopic realities that fill us with wonder and arouse in us the sensation of sharing in the intimacy of a cosmic revelation.

This study has sketched a history of the artistic processes of immersing the camera within terrestrial fluids, of the related investigations into infinitely complex physical phenomena beyond our senses and of the inventions of appropriate plastic forms whose heuristic power gives us access to sensitive, unprecedented and wonderful conceptions of the world and man. It remains to be completed by integrating the analysis of other films by the artists mentioned and extending it to the initiatives of many other experimental filmmakers, starting with the works of Peter Hutton, Chris Welsby, Philippe Cote or Anja Czioska. It has, however, established that filmmakers can enhance the functionalities of programmed optics, without having prior technical knowledge, without acting directly on the lens, nor making calculations in advance. The countless translucencies of terrestrial fluids offer unlimited optical and plastic resources, far more sophisticated than those of the glass lenses arranged within the cameras. While the filmmakers criticize these for their standardized optical principle of image formation, they nonetheless validate them as powerful observational instruments and make the most of their modulation transfer functions, their resolving powers and their ever more sophisticated transformation of light.

Notes

1. Jean Epstein, "Le cinéma du diable", in Nicole Brenez, Joël Daire and Cyril Neyrat (eds), *Jean Epstein, Écrits complets: L'Intelligence d'une machine, Le Cinéma du Diable et autres écrits, vol. 5, 1945–1951*, Paris, Independencia, 2014, p. 151: "Each of our senses is only capable of detecting a limited range of certain movements. On either side of the infrared and ultraviolet, there is only night, in which this machine for multiplying the rhythms of photon waves, this gearbox that is the retina, can no longer calculate. On either side of the infra-low and the super high frequencies,

there is only silence. Similarly, our mind is only tuned to conceive reality within a certain quantitative zone of spatio-temporal data. If there are too few or too many references on either side of these boundaries of reality, nothing can exist for us. Between these two extreme zones of incipient and evanescent reality, we can trace a curve where all the degrees of reality are located, which, first increasing then decreasing, vary first directly and then inversely in proportion to the number of references, dimensions and rhythms that a phenomenon accumulates." Unless otherwise noted, all translations from original French are by the author and translator.

2. Jean Epstein, "The Cinematograph Seen from Etna" [1926], trans. Stuart Liebman, in Sarah Keller and Jason N. Paul (ed.), *Jean Epstein. Critical Essays and New Translations*, Amsterdam, Amsterdam University Press, 2012, p. 292.
3. Jacques Perconte, "Bien plus fort que la Haute Définition", in Nicole Brenez and Bidhan Jacobs (eds), *Le Cinéma critique: De l'argentique au numérique, voies et formes de l'objection visuelle*, Paris, Publications de la Sorbonne, 2010, p. 233.
4. *Ibid.*, p. 234.
5. See Nicole Brenez's analysis of Mesguish's film in her "Poèmes argentiques contemporains: génie de l'instable", lecture given at the Colloquium *Lumière des Lumières*, Le Fresnoy, 2011, in *Turbulences vidéo*, no. 78, January 2013, pp. 62–63.
6. Jean Epstein, "Le cinématographe dans l'Archipel" [1928], in *Écrits sur le cinéma 1921–1953, Tome 1: 1921–1947*, Paris, Seghers, 1974, p. 197.
7. Robert Cahen, born in 1945, has been a major experimenter of filmic and electronic instruments since the early 1970s and digital ones from the 1980s. Indeed, in 1981, Robert Cahen used the very first digital camera in France at the Polytechnic School for *Artmatic* (1981, colour, sound, 4′15").
8. Robert Cahen quoted by Marco Maria Gazanno, "Robert Cahen et la matière", in *Robert Cahen s'installe*, exhibition catalogue, Sélestat, FRAC Alsace/Agence culturelle d'Alsace, 1998, p. 19.
9. According to Jacques Perconte's notes "Démarche – Madeira", 2014.
10. *Ibid.*
11. Introduced in 2011 for his "Upper Normandy" cycle, including *Moutiers, Wave cows, Impressions*, and his "Madeira" cycle, which

remained his preferred detection tool until 2014. From September 2014, Jacques Perconte uses the professional camera SONY PXW-X160 instead.

12. The modulation transfer function (MTF) describes the behaviour of the lens as a function of spatial frequencies and, consequently, the contrast ratio restored as a function of the fineness of the details of the detected phenomenon. The resolving power represents the limit of detail restitution at the highest frequencies. See Philippe Bellaïche, *Les Secrets de l'image vidéo, colorimétrie, éclairage, objectif caméra, signal vidéo, compression numérique, format d'enregistrement*, 6th edition, Paris, Eyrolles, 2011, pp. 133–138.
13. Ultra Low Dispersion.
14. Leighton Pierce, born in 1954, photographer, filmmaker, video artist, digital artist and teacher, has been the director of CalArts since July 2013.
15. Bill Viola, in Raymond Bellour, "La sculpture du temps: Entretien avec Bill Viola par Raymond Bellour", *Cahiers du Cinéma*, no. 379, January 1986, p. 38.
16. *Ibid.*, p. 40.
17. Bill Viola, in Jérôme Neutres, "L'art est un exercice spirituel: Conversation avec Bill Viola et Kira Perov", in *Bill Viola*, exhibition cat., Paris, Editions de la Réunion des musées nationaux – Grand Palais, 2014, p. 24.

PART III

A Decentred Cinema

CHAPTER 14

Lothar Baumgarten: Between Amazonian Myth and Animist Aesthetics

Lucas Murari

"*We plant trees and trees plant us, since we belong to each other and we must coexist*", Joseph Beuys.[1]

Thinking Beyond the Human

A fundamental pillar of the movements of ecological thought is the development of a perspective capable of surpassing anthropocentrism, that is, proposing a multi-species ethnography. Thus, the contemporary resurgence of the debate around animism reflects the need for new forms of thought capable of complexifying the understanding of a world divided and intersected by numerous socio-natural crises. The term animism comes from anthropology, and was introduced by Edward Tylor in his work *Primitive Culture* (1871) to refer to what he then considered the

L. Murari (✉)
Federal University of Rio de Janeiro, Rio de Janeiro, Brazil
e-mail: lucasmurari@gmail.com

E. Della Noce and L. Murari (eds.), *Expanded Nature*, Experimental Film and Artists' Moving Image,
https://doi.org/10.1007/978-3-031-70729-2_14

most "primitive" form of religion. It is therefore rooted in a rationalist project of modernity and cultural evolutionism, a current of anthropology that is considerably contested today. In recent decades, many authors such as Philippe Descola, Isabelle Stengers, Eduardo Viveiros de Castro, Tânia Stolze Lima, Tim Ingold, Graham Harvey, and Donna Haraway have sought to revive the embers of this discussion while updating the term,[2] not only as a religious belief but above all as an ontology. As an anthropological construct, the concept of animism designates a certain understanding of the world found in tribes located in the North and South of the American continent, Siberia, and certain parts of South-east Asia, without the term itself being used by the peoples in question. The debate around animism is manifested today in many disciplines within the humanities, arts, and sciences, significantly contributing to the deconstruction of certain epistemes rooted in Western thought.

At the end of the twentieth century, the North American art critic and historian Hal Foster observed the strong impact caused by ethnography in contemporary art and theory, an impact to which he gave the name "the ethnographic turn".[3] If, in the middle of the last century, anthropologists harboured a certain jealousy towards artists—such as with certain canons of ethnographic cinema exemplified by Gregory Bateson, Jean Rouch, Timothy Asch, and David MacDougall, all trained anthropologists—what the ethnographic turn entails is that now it is the artists who find themselves consumed by a desire for ethnography.

The activist artist now focuses on "the other", defined by their cultural and/or ethnic belonging rather than their socioeconomic status (like the exploited proletariat) as envisaged by an earlier moment of political art.[4] According to Foster, there are certain risks inherent in the artist posing as an ethnographer: he mentions the possibility of a new type of cultural patronage, where the artist acts as a "protector", but also of colonialist exploitation, where the artist becomes an authority figure. Beyond the artistic field, Foster considers this discussion fundamental for the social sciences as a whole.

Lothar Baumgarten: The Artist as Ethnographer

In the field of contemporary art, Hal Foster particularly focuses his attention on the works of Lothar Baumgarten,[5] whose artistic work relies on disciplines such as anthropology and ethnography. Through a series of visual devices, the German artist seeks to intertwine key concepts of Western thought, such as nature and culture, human and non-human, civilized and primitive, Europe and New World. It is notably through contact and conversation with Amerindian peoples that the avant-garde artist questions some of these parameters. Indeed, since the 1970s, Baumgarten has created various works among the Indigenous peoples of the Americas, aiming to blur certain binary principles through complex aesthetic elaborations, thus questioning the "truth" that resides in technical images and Western scientific reason.

Between 1977 and 1986, Baumgarten travelled to the Amazon region of Brazil and Venezuela six times. Between 1978 and 1980, he stayed there for nearly eighteen months, in the Upper Orinoco basin, between the Yanomamis of Kashorawë-theri and the Nyapetawë-theri. At the time, contacts between these tribes and Western civilization were rare. The artist reflects on this experience: "You cannot reflect your own society, unless you know a society that is remote from it. To know that society, you don't want to walk into it, get a Ph.D. and turn the page. You have to jump into the bushes, almost naked, as I did."[6]

His first feature film, *The Origin of the Night (Amazonian Cosmos)* [*Der Ursprung der Nacht (Amazonas Cosmos)*], which he made between 1973 and 1977 and shot in 16 mm, is based on the structural analysis by anthropologist Claude Lévi-Strauss of a myth of the Amazonian Tupi tribe about the origin of the cosmological alternation of night and day:

> In former times, night did not exist. It was daylight all the time. Night slept beneath the waters. Animals did not exist either, for things themselves had the power of speech.
>
> The daughter of the Great Snake had married an Indian, who was master of three faithful servants. 'Go away,' he said to them one day, 'for my wife refuses to sleep with me.' But it was not their presence which embarrassed the young woman. She wanted to make love only at night. She explained to her husband that her father held night prisoner, and that he should send his servants to fetch it.
>
> When they arrived in a canoe at the abode of the Great Snake, the latter handed them a tightly dosed nut of the tucuman (*Astrocarjum incimm*) palm, and told them not to open it under any pretext. The servants re-embarked and were soon surprised to hear a noise coming from the nut: ten, ten ...xi, like the sound of crickets and little toads which sing at night. One servant wanted to open the nut, but the others were opposed to the

idea. After a good deal of discussion and after they had travelled a long way from the Great Snake's abode, they eventually all assembled in the middle of the canoe where they lit a fire and melted the resin sealing the nut. At once night fell and all things that were in the forest changed into quadrupeds and birds; those in the river became ducks and fish. The basket turned into a jaguar, the fisherman and his canoe became a duck: the man's head acquired a beak; the canoe became the body, the oars the feet.

The darkness covering the world made it clear to the Snake's daughter what had happened. When the morning star appeared, she decided to separate night from day. With this end in view, she changed two balls of thread into birds - the cujubim and the inhambu [respectively of the cracidae and tinamidae families] and which sing at regular intervals during the night or to greet the dawn; in connection with these 'bird-clocks'.[7]

Lévi-Strauss's study of this myth is published in the second volume of the series *Mythologiques*, entitled *From Honey to Ashes*. The French anthropologist highlights three oppositions that form the framework of his reflection: (1) the division between day and night; 2) the opposition between conjunction and disjunction of the sexes; and 3) linguistic behaviour and non-linguistic behaviour. In Indigenous myths, we distinguish those of a nocturnal character from those of a diurnal resonance. In this latter category, the separation of the night ensures the reign of light; it is precisely to this type of myth that *The Origin of Night* belongs. For the Tupi tribe, time is relative, day and night are united by a relationship of reciprocal mediation, which consists in a balance of the two terms. The separation between day and night is not established through separated extremes, but as the result of regular alternations that unite them. In this story, the night is literally an apparatus of darkness, the orifice from which nocturnal and noisy animals spread—insects and amphibians—which are precisely those whose name designates the instruments of darkness in Europe: frog, toad, cricket, grasshopper, locust, etc. However, the contrast between day and night also implies differences in linguistic and non-linguistic behaviour: when the light of day still reigned, everything was endowed with speech, including animals and objects. But at the precise moment when night falls, everything becomes mute, and the voices of animals are reduced to inarticulate cries, yelps, meows, howls, chirps, roars, etc. Lévi-Strauss compares this distinction among many Indigenous myths and observes how these "moved from a spatial field to a temporal scale and, what is more, from the notion of absolute space to that of relative time".[8] The burnt world evokes fire and light; the rotten world, on the other hand, evokes darkness and night. The *Mythologiques* series is marked by an interpretation that is fundamental for modern and contemporary anthropology: the relationship between nature and culture. The very subtitle of the second volume indicates this: honey is to nature what ashes are to culture. Amerindian myths frequently recount stories from a time when humans and animals were not yet separated, a period of interspecies coexistence.

The Origin of the Night: Amazonian Cosmos

Baumgarten invokes the myth *The Origin of the Night* to express his own perspective of an animist Amazonian cosmology. His approach blurs the boundary between nature and culture, the realms of the sensible and the intelligible—both the film and the myth resist binary oppositions. Not having yet had access to this microcosm, the human being is absent from the feature film; likewise, there is no depiction of the characters described by the Tupis (the daughter of the great serpent, her husband, the three faithful servants). The artist is content to film, in an immersive way, the diversity of images and sounds offered by the forest—in short, the natural world (animals, plants, streams)—the space *par excellence* of fables and myths. The myth, in this sense, configures an experience of the ritual order, exposing an extraordinary and transfigured world through a story about the origin of things. Indigenous thought is approached by Baumgarten as a philosophical *praxis*, as an alternative ontology—the "metaphysics of others", according to the expression coined by Eduardo Viveiros de Castro[9]—which complicates our own perception of what is filmed by the camera. He combines the elements filmed to construct his own aesthetic:

> Though trapped in my Western thought patterns, I have always been interested in the "other"—the societies with out a state—and in historical and social cohesion, the great anthropological nexus, the unconscious, which maintains a common thought process in mankind of which it is barely aware, and determines their actions by means of the dialogue of myths. I am making an analytical effort to free the historically and socially based reality from the imposed myths cloaking it, and to raise it by rational criticism to a metaphorical level, to art. It is the desire to anchor an aesthetic dialectical praxis in the social and political conditions.[10]

The film begins with a completely black image, plunged into silence. It is the night, which envelops the start of the film, and which we will encounter again at its end. The work thus encapsulates the balance of the Tupi myth between day and night, completing a full cycle from night to day, and from day back to night. The opening of the film presents us with a constellation of words, gradually appearing on the screen until they almost entirely fill it. In all, there are 52 names, in both German and Tupi, designating animals, plants, trees, and insects, all native to South America. The words vanish and it is now the subtitle of the film that appears on the screen: *Amazonas Kosmos*. Again, in a similar process, the screen fills with words in German and Tupi. After these have also disappeared, the title of the film then takes their place: *The Origin of the Night* [*Der Ursprung der Nacht*].

The opening ends, and a female voice enters the scene. She tells us, in German, the Tupi myth eponymous with the title of the film. The text used is the one found in the work of Lévi-Strauss, *Mythologiques II. From Honey to Ashes*. The majority of the story is told to us in a voiceover, on a completely black image. Only a drawing breaks the darkness, in the middle of the myth, exposing us to the mysterious figure of a campfire, surrounded by bipedal insects and vegetation. The myth immerses us in the Indigenous imagination. The use of narration serves as a kind of prologue. While the audiovisual elements seem to suggest that the material was recorded in the Amazon rainforest, the homeland of the ethnic group in question, the last card of the film however clarifies: filmed in the Rhenish forests [*gedreht in den Rhein-Wäldern*]. The artist confuses us in our understanding of the represented space, inducing a fallacious approximation through the film's subtitle. The Amazon referred to here is, in fact, an artistic-intellectual construction, a speculation situated at the intersection of Indigenous mythology and European vision. The images and sounds that make up the work were recorded in southwest of Germany, near the city of Düsseldorf, more specifically in the Black Forest, a macroecosystem renowned for its fauna and flora, and possessing a rich biodiversity.

An Aesthetic of the Forest

Throughout the narrative, the forest reveals itself to be the main character of the film. The reference to the "Amazonian Cosmos" signals to us that this ecological system, a symbol of the environmental crisis, deserves particular attention on a global scale. The manifest interest in the different forms of life in the tropical jungle does not refer us to a distant country, but to the utopia of another life; it is the elegy of a world that is collapsing. The local specificity appears to be, in fact, only a kind of illusion if we only consider the Indigenous cosmology. In an interview with curator Christian Rattemayer, Baumgarten comments on this curiosity:

> I had an early interest in the cosmos as an entity. Long before the word 'ecology' started appearing in the features section of our newspapers, I had already become acquainted with various myths in the library at home and discovered in the taboos they divulged that a code of behavior was needed to govern exchanges between nature and culture.[11]

The first sequences following the opening show us an urban landscape: diffuse points of light in the darkness, cars driving at high speed, the sound of traffic, a street. And it is night that prevails. The following shots are also marked by total darkness. Lightning, heavy rain, and quick close-ups of amphibians mark the transition to the daytime period, which takes place entirely in the forest. From now on, the camera seems to move freely within it: trees, flowers, and a wide diversity of shrub vegetation; marshes, streams, raindrops; but also various animals, such as frogs, dragonflies, snakes, and herons. The artist relies on the observations of Joseph Beuys and concentrates many of his shots in one of the most unique spaces in Europe: the marshes, which according to Beuys are "the liveliest elements in the European landscape, not just from the point of view of flora, fauna, birds and animals, but as storing places of life, mystery and chemical change, preservers of ancient history".[12] We can also highlight the abundance and regularity of aquatic images present in the film. Baumgarten emphasizes their importance through saturated colours. This is one of the fundamental elements of the myth addressed by the work. The Tupi cosmology states that "night did not exist", that "it was daylight all the time".[13] And it is precisely at the bottom of the water that the night lay, a prisoner of the Great Serpent. This dimension is also important for other Indigenous ethnic groups, like the Bororos, who associate the bottom of the water with the *agié*, an aquatic creature of terrifying aspect. If water is very present throughout the *Mythologiques* series, it is because, according to Lévi-Strauss, it is a substance that permeates the myths of the entire continent. In these myths, Indigenous peoples distinguish between stagnant water and running water, also recognizing other variations, such as relatively high water/relatively low water (marsh), one being "pure" (alive) and the other "impure" (dead), due to its inability to flow. In South American myths, running water is considered the most powerful and effective, but also the most dangerous, as it is inhabited by spirits or directly related to them. According to Lévi-Strauss, it is the element that can break the fragile link that exists between a supernatural character and a human being.[14]

The film's framing within the forest oscillates between close-ups and wide shots; the image is either static or rhythmically fluid. The artist's experimentation is also realized in visual terms, through blurred shots or those filmed so closely that they border on abstraction. The artwork is sensitive to the plastic beauty of the world, exposing us to resplendent stagnant waters, moving clouds, frogs by the lake, or even animal eyes in extreme close-ups. The artist emphasizes the meticulous observation of local biodiversity. For this, an attentive eye will be able to detect that the environment depicted does not resemble the Amazon forest, showing no trace of jaguar, monkey, parrot, manilkara, orchid, lianas, or other common components found in the tropics and foreign to the European continent.

Contrary to a certain tradition of documentary cinema addressing the theme of ecology, the artist's objective here is not to explain what the forest is or how it functions through cinematic processes. The work rejects all didactic aspects, and the minimal information provided is intentionally ambiguous. The film follows an experimental path, focusing on the connection between nature and culture, the interaction between the Western gaze and Indigenous cosmology, and the blending of written and audiovisual languages. In some shots, remnants of human civilization appear: planes flying over trees, tables and chairs in the middle of the forest, a pan and a knife hanging from a tree. These elements function as traces, signalling the ambiguity of the coexistence of cultural objects and natural elements. For Lothar Baumgarten, nature and culture, as systems, do not operate in opposition, as if they were absolute divisions. The work he created before *The Origin of the Night* is precisely named *Culture-Nature, Manipulated Reality* (1968–1972) [*Kultur-Natur, Manipulierte Realität]*. It is a series of images that propose a reflection on naturalist iconography and Western cultural representation. The hyphen in the title suggests both hybridity and continuity.

Final Considerations: Multinaturalist Cosmologies

The film *The Origin of the Night* is a Western interpretation of the Amazonian Tupi myth and Amerindian cosmology. The images and sounds take us back to the time of origins, prior to metamorphoses, when there was still no distinction between humans and animals. Baumgarten films the forest in search of this animist aesthetic, without establishing a division between what is commonly called Nature and Culture. This is what the artist seems to indicate himself when he comments in an essay titled *Status Quo* on the spiritual and intuitive force that marks his work:

> I am interested in the metaphysical force that things are imbued with. The intenseness of their presence, the "being" of material and color, the flavor of a thing, the relation of proportion to form, and then again mass and weight, space and time. My involvement is with the dynamic energies of this rhythm. It is a continuous attempt to resolve the contradiction of spirit and matter through the imagination. It is the self-same journey to the origins, back to the truth: the dream of creating the world all over again within oneself. It is the yearning, the struggle for a grammar that would not run counter to the laws of natural order, yet could serve to protect the independence of the artist from nature. A process of discovery where instrument and idea are inseparably fused and of equal weight.[15]

The ethnographic approach adopted by Baumgarten relies on the sensory perception of the filmed forest environment, highlighting the multiplicity and richness of different forms of life, without privileging any particular species. The languages of the Amazonian peoples have been constructed in close relation with the forest, so many of the objects in their daily life incorporate figurative elements, including zoomorphic and anthropomorphic traits. This is an animistic vision of the world: everything that exists in the cosmos can be considered a subject. The artist deconstructs the iron cage of epistemology that imprisons modern thought around its own foundations, challenging the perspective of the social world as an exclusively human world. The end of the feature film is dedicated to the return of the night, represented by the moon and darkness, as well as by images of a campfire, a key element of this passage and the Tupi myth. A male voice then declares: "it will never be day again", and proceeds to cite various names of South American Indigenous tribes: Jivaro, Tukano, Tupinamba, etc. This finale marks the original cosmological alternation of night and day, so important for the myth.

Notes

1. Joseph Beuys, in Lucrezia de Domizi Durini, *The Felt Hat: Joseph Beuys, a Life Told*, Milan, Charta, 1997, p. 66.
2. It is important to highlight some anthologies that have been published in recent years on the various discussions on animism: Graham Harvey (ed.), *The Handbook of Contemporary Animism*, Durham, Acumen, 2013; Anselm Franke (ed.), *Animism*, Berlin, Sternberg Press, 2010; Teresa Castro, Perig Pitrou and Marie Rebecchi (eds), *Puissance du vegetal et cinema animiste: la vitalité révélée par la technique*, Dijon, Les Presses du réel, 2020.
3. See Hal Foster, *The Return of the Real: The Avant-Garde at the End of the Century*, Cambridge, The MIT Press, 1996.
4. Hal Foster refers to Walter Benjamin's speech "The Author as Producer", delivered on April 27, 1934 at the Institute for the Study of Fascism, in Paris.
5. Lothar Baumgarten studied in 1968 at the Academy of Fine Arts in Karlsruhe (Staatliche Akademie der Bildenden Künste Karlsruhe) and then, from 1969 to 1971, at the Academy of Arts in Düsseldorf (Staatliche Kunstakademie Düsseldorf), where he was a student of Joseph Beuys. Beuys advocated for the idea of a certain shamanism in contemporary culture, considering the artist himself as a shaman, as a mode of surfacing a sensitive perception of the world.
6. John Russell, "Art View: Lothar Baumgarten's Discreet Provocations", New York, *The New York Times Archives*, 1988.
7. Claude Lévi-Strauss, *From Honey to Ashes: Introduction to a Science of Mythology/Volume 2*, trans. John and Doreen Weightman, Chicago, University of Chicago Press, 1973, p. 416.
8. *Ibid.*, p. 420.
9. See Eduardo Viveiros de Castro, *Cannibal Metaphysics*, trans. Peter Skafish, Minneapolis, University of Minnesota Press, 2014.
10. Lothar Baumgarten, "*Status Quo*", *Artforum International*, New York, Vol. 26, No. 7, 1988, p. 108.
11. Lothar Baumgarten, "Status Quo", art. cit , p. 108.

12. Christian Rattemeyer, "Lothar Baumgarten Conversation", in Kaira Cabanãs (ed.), *Lothar Baumgarten: Seven sounds, seven circles*, Bregenz, Kunsthaus Bregenz, 2009, p. 136.
13. Joseph Beuys as quoted in Caroline Tisdall, *Joseph Beuys*, New York, Solomon R. Guggenheim Museum, 1979, p. 39.
14. Claude Lévi-Strauss, *From Honey to Ashes: Introduction to a Science of Mythology/Volume 2*, *op. cit.*, p. 416.
15. Lothar Baumgarten, "*Status Quo*", *art. cit.*, p. 108.

CHAPTER 15

Reanimating Nature: In Praise of an Animist Cinema in the Age of Ecological Crisis

Teresa Castro

André Gay's evocation of the new dazzling spectacle of moving images in the pages of the *Revue générale des sciences pures et appliquées* (1895) contains a detailed description of one of the very first Lumière views, showing two blacksmiths at work.[1] Testifying to the early spectator's fascination for fleeting phenomena—among which the famous trembling of the leaves in *Repas de bébé* [*Baby's Meal*] (Lumière Brothers, 1895)—, he lingers on the cloud of steam emanating from the water on contact with the burning iron "and which a gust of wind suddenly sweeps away all at once".[2] And he concludes, "in the words of Fontenelle, nature itself was caught in the act".[3]

Whatever may be understood by "nature", the film medium has captured it from its beginnings with an accuracy and a vitality that has felt new. Sometimes (as in *Repas de bébé*), nature imposed itself on the eyes of spectators from the background where it is traditionally confined.

T. Castro (✉)
Sorbonne Nouvelle University, Paris, France
e-mail: teresa.castro@sorbonne-nouvelle.fr

E. Della Noce and L. Murari (eds.), *Expanded Nature*, Experimental Film and Artists' Moving Image,
https://doi.org/10.1007/978-3-031-70729-2_15

Very early on, however, it became the subject in its own right of certain films, whether in the guise of "landscape" or, more broadly, of the "natural world". The latter is made up of beings and phenomena on which cinema confers a singularly moving and empathetic power. At other times, the cinematographic apparatus became an instrument through which "nature" is caught unaware in its most discreet actions. Indeed, cinema (especially so-called scientific cinema) has expanded the curious and attentive gaze to which Bernard de Fontenelle refers when he wrote, in 1724, the words that Gay recalled almost two centuries later: "In vain had nature hidden itself in so deep and inaccessible places to work on the vegetation of stones; it was, so to speak, caught in the act by such hardy observers".[4]

Stimulated by the desire to explore the potentialities of the medium and freed from the constraints imposed by the industrial production system and its normative gaze, experimental cinema has often made "nature" the subject of its experiments. To mention but two examples related to aspects mentioned above, with *Un vent léger dans le feuillage* (1994) the French filmmaker Martine Rousset pushes to its extremes the "metaphysics of leaves" initiated by the Lumière brothers in *Repas de bébé*,[5] transforming the spectacle of their movement into an experience of "chlorophyll's fluorescent vibration"[6] (Rousset films a few leaves in a motionless close-up shot) (Fig. 15.1).

Meanwhile, British filmmaker Charlotte Pryce pursues the long tradition initiated by 18th-century botanists (those whom Fontenelle praises for their solicitude), by delving into the forest floor in *Discoveries on the Forest Floor* (2006).[7] The subject of her three-part film (*Burnt Umber/ pale ochre/ Burnt Umber*; *The Talk of Lichen on a Lonely Day*; *Those Whose Attachment to the Earth is but Tentative*) is none other than the "natural world". By immersing us in this other world, *Discoveries on the Forest Floor* also reconnects with scientific imagery, drawing on illustrations from botanical works and mobilizing time-lapse photography (Fig. 15.2). The latter's ability to compress time and thus reveal phenomena invisible to the human eye has greatly interested scientific cinema, the experimental nature of which should not be forgotten.[8]

In the following paragraphs, I would like to draw attention to the way some films said to be "experimental" (including *Un vent léger dans le feuillage* and *Discoveries on the Forest Floor*) invite us to experience nature in ways that highlight its *vitality*. By focussing on this aspect, I wish to propose that film has the ability to reanimate nature. Note that I use the word *reanimate* and not animate. This implies shifting the debate on the

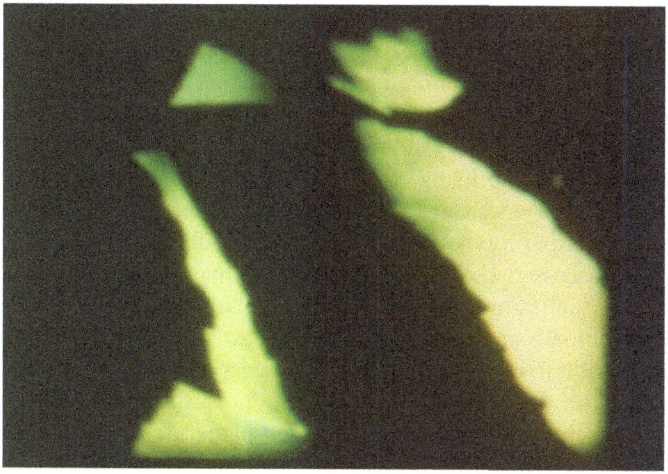

Fig. 15.1 Marine Rousset, *Un vent léger dans le feuillage* (1994) © All rights reserved by the artist/Courtesy of Light Cone

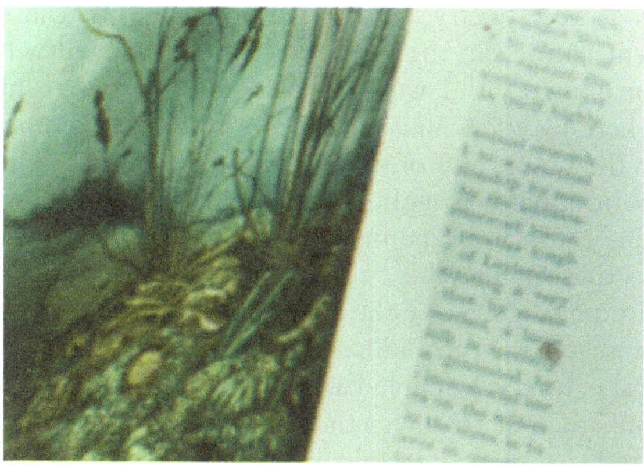

Fig. 15.2 Charlotte Pryce, *Discoveries on the Forest Floor* (2006) © All rights reserved by the artist/Courtesy of Light Cone

movement of images towards vitality, even animism, but also revisiting the question of what we understand "nature" to be. Indeed, to suggest that film is capable of reanimating nature means, on the one hand, abandoning a certain conception of the latter and, on the other, admitting that the former can shift the thresholds of attention and increase the sensitivity of its spectators to the living. Ultimately, my point here is to argue that film—in particular a certain type of cinema, entertaining a special relationship with experimentation and which I would describe as "animist" for reasons discussed below—can help us re-imagine nature, understood as the conceptual framework within which relations between humans and more-than-humans are negotiated.[9] This point seems all the more important as we are currently going through an ecological crisis which forces us to reconsider our relationships with the more-than-humans around us and upon which our survival on a damaged planet depends. As Australian philosopher Val Plumwood observed in 2007 (her writings permeate my reflection), "So the big question is: Can we think differently?"[10] Originally titled "Nature in the Active Voice", but translated into French as *Réanimer la nature* (*Reanimating Nature*), this short but incisive text briefly looks at how poetry and literature manage to reanimate matter and to present nature "in the active voice", i.e., imbued with a creative and intentional agency.[11] I would like to add cinema to this equation, represented here by a few films which, due to their experimental nature, dare to push the boundaries of visual exploration of the world. "Tracking the invisible, bringing it to light, giving it cinematic form, these are the stakes of avant-garde and experimental cinema", writes Christian Lebrat, reminding us that the invisible "can be understood in different ways".[12] The creativity and agency of the more-than-human world are part of this invisible that has been hidden from us on the basis of a doctrine that conceives nature as a simple backdrop for life, or as an inexhaustible and inanimate stock of resources that must be domesticated, organized, dominated, reduced to figures, calculated and above all managed.

Reanimating Nature

According to Edgar Morin, "the Lumière cinematograph … *reanimated*, without making us conscious of it, an animist or vitalist sensibility".[13] Watching *Un vent léger dans le feuillage*, you are struck by the vigour of the sparkling foliage, like a beating heart of another kind. Similarly, under the litter of dead leaves explored by Charlotte Pryce's camera, life springs from everywhere. Vitality here is a matter of those expressive resources specific to film: it's a question of framing, of scale, length and vibration of the shot, of its chromatic pulsation (*Un vent léger dans le feuillage*); it's an issue of speed manipulation, of radical transformation of scales, of flicker, contrasts and film stock envisaged as both a plastic and living material, itself immersed in the undergrowth evoked by the film (*Discoveries on the Forest Floor*). I insist again on the use of time lapse and slow motion: not only do they disclose the invisible, but they also reveal and generate intentionalities, creativities scattered among the mushrooms and plants of the forest floor.

To *reanimate*, and not to animate, nature—to revive, give life and vitality back to those who have lost them, those from whom such qualities were taken away. Viewed as a reality external to humans and their history, "nature" is a conceptual invention of the modern West which, driven by the rationalist enterprise and the development of the capitalist system, projected it as something universal and homogeneous, distinct from the multiplicity of cultures and even opposite to Culture.[14] Before the crucial works of Philippe Descola and Bruno Latour popularized (at least in the French-speaking world) the notion of the "great divide"—the nature–culture split—several ecofeminist historians and philosophers (among them Plumwood) had interrogated the mechanisms and historical reasons that led to the instrumentalization (and feminization) of nature in the West. The goal of these theorists was, and still is, to imagine more harmonious relationships with the Earth and more-than-human beings.

In *The Ecological Crisis of Reason* (2002),[15] Plumwood argues that the narrative that aligns the heroic supremacy of reason with human exceptionalism and superiority is at the root of the current ecological crisis. This narrative is not only the origin of human/nature, culture/nature, reason/nature dualisms, etc., but it also feeds into the suicidal (and murderous) fantasy of our ecological invulnerability. As a philosopher and ecofeminist, Plumwood does not advocate for some absurd chimera such as the "end of reason": the problem is not reason itself, but the reductionism (or reductionist materialism) of a certain type of rationalism and its destructive dimension. Exposing the inherent contradictions in the devastating and exterminatory irrationality of this malignant rationalism is only a first step for Plumwood: her goal is to lay the foundations of a more responsible rationality, aware of our generalized interdependence. While she draws inspiration from other cosmogonies (and in particular from Australian Aboriginal thought), Plumwood does not see them as a miraculous solution either. In other words, the question is not to become animist (or totemist or analogist)—even if that were possible for the naturalist beings that we are.[16] On the contrary, Plumwood wants to lay the foundations of an ecological rationality that is also an ethics of *care*, of caring for nature and more-than-humans.[17]

In Praise of an Animist Cinema

Mostly known for her sublime *Bouquets* series and her images of flowers, gardens and landscapes, in *Tartarughe d'acqua* (2016) the French filmmaker Rose Lowder is interested in freshwater turtles. Piled up in a small basin, they bask in the sun. Lowder's camera films them with kindness, documenting their quest for a place in the sun or the precarious balance of their acrobatic assemblages. The shots stretch themselves (and warm up): we experience something like what we imagine to be a "turtle space–time", a form of "taking time" that has nothing to do with lethargy. If the *Bouquets* or a film like *Coquelicots* (2000) are formidable perceptual adventures making nature vibrate with the help of a Bolex, *Tartarughe d'acqua* is a film about attention and sensibility towards the living. The turtles gain ontological consistency: they are. Here lies the film's astonishing vitality.

Establishing an ecological rationality—a rationality more sensitive and attentive to more-than-humans, more considerate towards the Earth, a rationality where the centrality and exceptionality of the *anthropos* is nuanced—is not an easy task. It requires us to free ourselves from some of our hierarchies of thought: those that justify the blind exploitation of "nature" and more-than-humans. As Plumwood suggests, literary, artistic and cultural forms can help us in this complicated task, the keystone of which is attention. As French philosopher Baptiste Morizot explains, the ecological crisis is above all a crisis of political (in)attention to forms of life and ecosystems.[18] In this context, we need to "learn to pay attention anew", to "reconstruct paths of sensibility to living beings" and "repopulate" the world—in the manner (among others) of *Un vent léger dans le feuillage*, *Discoveries on the Forest Floor* and *Tartarughe d'acqua*. To put it another way, in our age of attention capitalism, attention must be conceptualized not in terms of an economy, but of an ethics. Furthermore, such an ethics must be situated at the level of attentive perception rather than intimate feeling.[19] It is in relation to these topical issues that I would like to think about an animist cinema.

This expression, "animist cinema", is certainly intriguing. It evokes first and foremost a chronologically locatable discourse (1920–1950), whose history can be retraced. In this context, "animist cinema" refers to a series of debates about the latent or manifest powers inherent to filmic images (discussions notably illustrated by the theoretical writings of French filmmaker Jean Epstein, but not only: film's animism is a more widespread critical topos than one might think). However, I would like to detach myself from this historical interpretation and speculate on what an animist cinema could mean today, at a time of unprecedented ecological crisis. This qualifier may be surprising: when we talk about animism, we tend to think with anthropology. Thanks, in particular, to the very different works of several anthropologists (including Nurit Bird-David, Philippe Descola, Tim Ingold, Nastassja Martin and Eduardo Viveiros de Castro), animism is today recognized as an ontology among other ontologies. In other words, animism is a particular way of composing the world, characterized, according to Descola's minimal definition, by "the attribution by humans to nonhumans of an interiority identical to their own".[20] However, it is not with anthropology that I wish to think about animist cinema, but with Plumwood's ***philosophical animism***. Following Diane Linder, the latter is based on an "enriched materialism" (and has nothing to do with New Age spiritualism, as some sceptical readers might suspect), that is to say on the attempt to revalue beings and the so-called material world.[21] According to reductionist materialism, matter is devoid of "mind", agency, creativity. Plumwood wishes to go beyond the reified separation between matter (body) and mind, and to think differently about nature, the more-than-human world. Discerning creativity, intentionality, sentience, even intelligence in these places and beings is key—in particular because these notions were long anchored to the *anthropos*, something that science itself is now calling into question. The filmic animism that I have in mind refers to this project of animating and intentionalizing; formally, it is distinguished by a set of highly variable qualities that characterize certain representations of matter and of the living world: "chlorophyll's fluorescent vibration" in Rousset's film, Pryce's living film strip, Lowder's poikilothermic shots.[22] But because reanimating the world is also "remaking ourselves", animist cinema potentially works towards the establishment of a different sensibility.[23] In this sense, it renews the wonder with which the "natural world" inspired cinema's early 20th-century spectators. As the actress Georgette Leblanc wrote in a text from 1919, one goes to the cinema to catch a glimpse of "life in all its forms. Beings, beasts, insects. Trees, plants, rivers, sea, rain or sun. The earth and all the elements."[24]

The point of animistic cinema is neither to naively praise animism as some kind of miraculous panacea, nor to perpetuate a dualistic opposition between a nostalgic primitivism disguised as animism and a modernity haunted by its own death drives. We must move away from opposition, binarity and dualism. But we also need to free the notion of animism from the mistrust that surrounds it because, as Canadian anthropologist Natasha Myers aptly writes, "the very taboos against animism and anthropomorphism are grounded in colonial imaginations of nature and culture, and [the] disavowal of nonhuman sentience is intimately bound up in colonial projects that have taken shape under the guise of the ecological sciences".[25] To a certain extent, the aim is also to free "cinema", or at least some of its more ambivalent images, from the essentialist and sometimes punitive discourse that inevitably ties it to the objectification and domination of the world and nature. It is certain that "cinema" is not only one of the inspired toilers of anthropocentrism, but also one of the troopers at the service of this "fundamental event of modernity" which is, according to Heidegger, "the conquest of the world as picture".[26] Paradoxically, however, as experimental cinema recalls, in the guise of prodigal child of techno-scientific modernity, what (animist) cinema can do (the power of images is a virtuality that can be actualized or not) is to help us shift the thresholds of attention, to forge a new sensibility, to work towards an ecological rationality.

Thanks to its specific powers, cinema has enabled us (and still enables us) to reanimate nature and restore our wonder towards the world. The role of experimental cinema in this process has been crucial, not least because "experimental film considers cinema not in terms of its uses, but in terms of its capacities [*puissances*]".[27] According to Plumwood, it is this ability to marvel at the world, to delight in the rustling of the leaves of a tree, that modernity has stolen from us. This sense of wonder, which experimental cinema has the capacity to restore, is more necessary than ever to us, particularly when it comes to rediscovering the intentionality of the more-than-human beings that surround us.

Notes

1. André Gay, "Actualités scientifiques et industrielles: Le Cinématographe de MM. Auguste et Louis Lumière", *Revue générale*

des sciences pures et appliquées, no. 14, 30 July 1895, pp. 633–636. Regarding the Lumière view in question, it is an as yet unfound version of view no. 51, *Les Forgerons* [*The Blacksmiths*], 1895. See: https://catalogue-lumiere.com/forgerons. [accessed 14 June 2021]. Unless otherwise noted, all translations from the original French by the author and translator.
2. *Ibid*, p. 636.
3. *Ibid.* Gay is referring to the author and scientist Bernard de Fontenelle (1657–1757).
4. B. de Fontenelle, "Éloge de M. Tournefort", in *Œuvres diverses*, vol. 3, Paris, 1724, p. 187.
5. I take the expression "metaphysics of leaves" from the study by François Albera, "Le Cinématographe dans le mouvement: une métaphysique des feuilles", *1895. Revue de l'association française sur l'histoire du cinéma*, 2019, no. 87, pp. 34–63.
6. Nicole Brenez, "Couleur critique (Expériences chromatiques dans le cinéma contemporain)", *in* J. Aumont (ed.), *La Couleur en cinéma*, Paris/Milan, Cinémathèque française / Mazzotta, 1995, p. 171.
7. Although Pryce's approach seems to me to also evoke that of the "hardy observers" who were busy examining "the vegetation of stones" (see again the quote from Fontenelle), the title of her film refers to a little-known sub-genre of still lifes, known as *sottobosco* (undergrowth in Italian). As in the artist's film, the subject of these paintings, inspired by the scientific discoveries of their time (the seventeenth century), is the flora and fauna of the forest floor.
8. The time lapse was first used for scientific experiments, as evidenced by four cinematographic studies on the movement of plants filmed by German scientist William Pfeffer between 1898 and 1900. See: https://av.tib.eu/media/12258. [consulted on 13 June 2021].
9. I choose here this expression, "more-than-human", to avoid the hierarchical opposition suggested by the pair "human" and "non-human".
10. Val Plumwood, "Nature in the Active Voice", *Australian Humanities Review*, issue 46, May 2009, p. 122.
11. On the choice to translate "Nature in the active voice" as *Réanimer la nature*, see the introduction by Diane Linder to Plumwood's work: Diane Linder, "Rendre la part animée à la nature", *in* Val

Plumwood, *Réanimer la nature* [2009], trans. Laurent Bury, Paris, PUF, 2020, pp. 17–18.

12. Christian Lebrat, "Invisible à l'œil nu", in N. Brenez and Ch. Lebrat (ed.), *Jeune, dure et pure ! Une histoire du cinéma d'avant-garde et expérimental en France*, Paris/Milan, Cinémathèque française / Mazotta, 2001, p. 23.
13. Edgar Morin, *The Cinema, or The Imaginary Man* [1956], trans. Lorraine Mortimer, Minneapolis, University of Minnesota Press, 2005, p. 65.
14. See, among others, Philippe Descola, *Beyond Nature and Culture* [2005], trans. Janet Lloyd, Chicago, University of Chicago Press, 2013. In the context of his anthropology of nature, Descola observes that in certain societies the concepts of "nature" and "culture" make no sense. On the links between the invention of nature, primitive accumulation and the development of capitalism, see Carolyn Merchant, *The Death of Nature: Women, Ecology and the Scientific Revolution*, San Francisco, Harper and Row, 1983.
15. V. Plumwood, *Environmental Culture: The Ecological Crisis of Reason*, London/New York, Routledge, 2002. See also her *Feminism and the Mastery of Nature*, London/New York, Routledge, 1993.
16. I am referring here to the ontologies identified by Philippe Descola: animism, totemism and analogism (Western naturalism being a fourth way of composing the world). See Philippe Descola, *Beyond Nature and Culture, op. cit.* As Claude Janin and Anne-Christine Taylor point out, according to this new anthropological understanding of the term, animism (and the three other ontologies) constitutes "a mode of cognitive functioning" in Claude Janin, Anne-Christine Taylor (ed.) "Introduction", *L'Animisme parmi nous: Colloque de la SPP*, Paris, PUF, 2015, p. 23. In proposing to establish an ecological rationality, Plumwood is not aiming for a radical change in cognitive functioning, but rather an adaptation of Western rationality to the challenges posed by the ecological crisis.
17. "Care" in the sense of Berenice Fisher and Joan Tronto: the ethics of *care* as "a species activity that includes everything that we do to maintain, continue, and repair our 'world' [our bodies, our selves and our environment, all of which we seek to interweave in a complex, life-sustaining web]". For a more detailed definition, see B. Fischer, J. Tronto, "Towards a Feminist Theory of Caring", in

E. K. Abel, M. K. Nelson, (ed.), *Circles of Care: Work and Identity in Women's Lives*, New York, State University of New York Press, 1990, p. 40.
18. Baptiste Morizot, *Ways of Being Alive* [2020], trans. Andrew Brown, Cambridge, Polity Press, 2022.
19. On capitalism and the economy of attention, see Yves Citton (ed.), *L'économie de l'attention. Nouvel horizon du capitalisme ?*, Paris, La Découverte, 2014.
20. Philippe Descola, *Beyond Nature and Culture*, *op. cit.*, p. 129.
21. Diane Linder, "De la nécessité d'une philosophie animiste", *in* V. Plumwood, *Réanimer la nature, op. cit.*, p. 75.
22. The turtle is a poikilothermic animal, that is, an animal whose body temperature varies with that of their environment. With the idea of a "poikilothermic shot", I am referring to the already mentioned feeling that the shots are warming up in Lowder's film.
23. V. Plumwood, "Nature in the Active Voice", *op. cit.*, p. 123.
24. See Georgette Leblanc, "Propos sur le cinéma", *Mercure de France*, no. 514, 16 November 1919.
25. Natasha Myers, "Ungrid-able Ecologies: Decolonizing the Ecological Sensorium in a 10,000 year-old NaturalCultural Happening", *Catalyst: Feminism, Theory, Technoscience*, vol. 3, no. 2, 2017, p. 7.
26. Martin Heidegger, "The Age of the World Picture" [1938], *Off the Beaten Track* [1950], trans. Julian Young and Kenneth Hayes, Cambridge, Cambridge University Press, 2002, p. 123.
27. Nicole Brenez, "L'Atlantide", in N. Brenez and Ch. Lebrat (eds), *Jeune, dure et pure ! Une histoire du cinéma d'avant-garde et expérimental en France, op. cit.*, pp. 17–18.

CHAPTER 16

Shamanic Materialism: 77 Theses on the Audiovisual

Colectivo Los Ingrávidos

Thesis 1: In political cinema of agitation there is a juxtaposition of the old and the new that entails an audiovisual practice in the form of aberration. In this cinema, agitation no longer emerges from new awareness or calls for mass mobilization, but rather consists of putting everything in a trance, including the camera itself, relating the different instances of violence to each other, so that the trance, the crisis or the aberration determines a constructivity that, intervened from the real, produces collective audiovisuals of the *Missing People*.

Thesis 2: A political cinema of agitation whose condition of possibility is the ravages of neoliberal capitalism and leads, in the audiovisual realm, to a double form of aberration: the sinister and estrangement.

Thesis 3: The exacerbation of the sinister leads to the intolerable. The vision, presence and insistence of the intolerable opens up a horizon that enables the emergence and autonomy of the Missing People.

C. L. Ingrávidos (✉)
Tehuacán, Mexico
e-mail: colectivolosingravidos@gmail.com

E. Della Noce and L. Murari (eds.), *Expanded Nature*, Experimental Film and Artists' Moving Image,
https://doi.org/10.1007/978-3-031-70729-2_16

Thesis 4: In the political cinema of agitation there are three forms of the People: The Supposed People, the Missing People and the Population.[1]

Thesis 5: The political cinema of agitation leads to an autonomy of images. Autonomy is not separatism, nor is it the taking over or occupation of the State, nor is it the totalitarian productivity of images, but rather the creation and appropriation of parallel sensory, perceptual, poetic and speculative structures within existing structures.

Thesis 6: One of the forms of the political cinema of agitation is the cinema of appropriation, which implies an aberrant sensory practice. The liturgical dimension of this aberration is burial (funerary trance).

Thesis 7: The burial of images and sounds is the kinetic trance of the appropriated audiovisual document.

Thesis 8: All such appropriation mobilizes and vindicates the opacity of the document.

Thesis 9: The opacity of the document is a sensory quality that activates the Burial-Exhumation dialectic.

Thesis 10: The Burial-Exhumation dialectic is the iconoclastic and materialist contradiction of appropriation cinema.

Thesis 11: Burial and Exhumation entail the testimonial dimension of appropriation.

Thesis 12: The sinister rapture of appropriation is shamanic possession, conversion and bonding.[2]

Thesis 13: Shamanic *possession* is the sensory outburst of the political cinema of agitation.

Thesis 14: Shamanic *conversion* is the hyperkinetic stasis of the cinema of agitation.

Thesis 15: Shamanic *bonding* involves the vindication of the irrational character of testimony as well as its supernatural drift.

Thesis 16: The perception of a testimony as supernatural, dream-like, messianic, excessive, irrational and insane is what enables a sensorial appropriation and *connection* by the document.

Thesis 17: The sensory appropriation of the audiovisual document presupposes an Aesthetics of Trance as a condition.

Thesis 18: In the Aesthetics of Trance, the immediate data of sensation induces an agitation in the perception of the documents.

Thesis 19: The mediate states of perception mobilize an appropriation of the documents.

Thesis 20: Agitation, appropriation and superstition are the impure practices of the Aesthetics of Trance.[3]

Thesis 21: The Aesthetics of Trance makes possible the semiotic collision and internality of materialism and shamanism.

Thesis 22: Materialism is the empirical, contingent, circumstantial, sensory and concrete dimension of objects, documents, archives, remains and elements.[4]

Thesis 23: Shamanism vindicates the suggestion, the omen, trance states, the spell and the superstition on the basis of objects and testimonies.

Thesis 24: Shamanic *materialism* proceeds by fragments, ruptures, loops, clusters, modules, drifts, ascents, descents, series, spirals, vortexes, pulses, rhythms, entropy, negentropy, hypostasis, collisions, linkages, aberrations, folds, burials, unearthings and everything which reveals intermittent constellations and from whose magma figures, agitations, forms, structures, processes, relationships, perceptions, speculations and sensations emerge. Shamanic materialism is a Mesoamerican spell unleashed.[5]

Thesis 25: Shamanic materialism makes simultaneous perception possible according to two incompatible, paradoxical and irreconcilable perspectives.

Thesis 26: Incompatible perspectives are, at bottom, unreconciled perspectives.

Thesis 27: The historical a priori of Shamanic materialism is non-reconciliation.

Thesis 28: Shamanic materialism asserts an eventual perspectivism of History.

Thesis 29: Non-reconciliation is the transhistoric core of History.

Thesis 30: History is a grave for the eye. An overdetermination for the ear.

Thesis 31: The eye and the ear are attractors of contingencies.

Thesis 32: Materialism entails a prodigious accumulation of contingencies.

Thesis 33: Any contingency can be appropriated, organized and reformulated as a multinaturalist circumstance.[6]

Thesis 34: Every multinaturalist circumstance is an immediately relational multiplicity.

Thesis 35: What exists in multinature[7] are not self-identical entities perceived differently, but immediately relational multiplicities.

Thesis 36: Shamanic possession is the sensory outburst of the political cinema of agitation in an environment of immediately relational multiplicities.

Thesis 37: Cinema-trance implies the ability to see simultaneously according to incompatible, paradoxical and irreconcilable perspectives.

Thesis 38: The cinema-trance of agitation induces a superposition of heterogeneous states as well as the explosion in the beings of qualitative multiplicity typical of myth.

Thesis 39: Myth consists of a record of the movement of updating the present state of things from a precosmological, prehistoric and ancestral condition.

Thesis 40: Every mythical metamorphosis entails an eventual trance of contingencies.

Thesis 41: Myth is a multinaturalist hyperimage in motion.

Thesis 42: The determination and dialectical organization of the collective contents of contingent images constitutes and displays a hyperimage.

Thesis 43: Every hyperimage expresses an epoch of the image of the world as well as a system of current relational images.

Thesis 44: Cinema is a dialectical hyperimage of motion.[8]

Thesis 45: TV is a media-relational hyperimage of culture.

Thesis 46: The Internet is a relational hyperimage of control devices.

Thesis 47: The pandemic is a metabolic overdetermination of systems of images.

Thesis 48: All metabolic overdetermination implies cognitive fracking as a practice of digital reason.

Thesis 49: Cognitive[9] fracking[10] is a productivist overdetermination that integrates and enables the relational hyperimage of control devices.

Thesis 50: Advertising, digital interaction and algorithmic processing of data instrumentalize the productivity of contingent reason.

Thesis 51: Every contingent reason enables and supports the factual, productivist and digital vector of reason.

Thesis 52: Productivist Reason implies an autophagic energy automatism.

Thesis 53: rEason is the productivist and autophagic nucleus of systems of images.

Thesis 54: reAson is the radical and progressive affirmation of the evidence of its reality.

Thesis 55: reaSon in the Anthropocene develops and affirms an energetic hyperimage of the world.[11]

Thesis 56: reasoN has only one way of explaining what does not emerge from it, and that is to reduce it to nothing.

Thesis 57: The reduction to nothingness, as a practice of reason, implies an eidetic contraction and dispersion of beings.

Thesis 58: Every eidetic reduction opens paths to extinction.

Thesis 59: The instrumentalization of eidetic reduction leads to precariousness as an existential condition in the epoch of reason in the Anthropocene.

Thesis 60: The precariousness of the images is a condition of possibility of audiovisual appropriation.[12]

Thesis 61: Extinction, exhaustion, entropy, wear, fatigue and destruction make up a possible ecological correlate of the directive images of contingent reason.

Thesis 62: The directive images of contingent reason imply three relational systems and processes of organizational internality: *the world* that *produces* reason in the Anthropocene, *the limits postulated* by critical reason and *the nature* that *links* speculative reason.

Thesis 63: *The world, limits and nature* constitute the guiding images of contingent reason.

Thesis 64: The constructivist differentiality of contingent reason according to its *connection* with each of these guiding images generates: (1) a productivist and objectifying reason in relation to the *world* (capitalist realism); (2) a critical and reflective reason in relation to *the limits* (ecological realism); and (3) a speculative reason in relation to *nature* (speculative realism).

Thesis 65: The first relational process of organizational internality (objectifying reason) implies a deterministic, scientistic, objectual, realistic, cognitive, epistemic production, as well as the assertion of the ontological one-sidedness of the world.

Thesis 66: The objectifying perspective of reason in the Anthropocene, in its productivist relationship with the world, presupposes a pre-existing and inherent structure to nature that enables the objectivity of the world.[13]

Thesis 67: Contrary to objectifying reason, the relational process of critical reason (ecological realism) presupposes the radical and irreducible exteriority and objectivity of nature on which basis the necessary limits for the destructive activity of reason in the Anthropocene will be determined.

Thesis 68: The reflective perspective of critical reason, in its ecological relationship with the world, implies the establishment of limits for the productivism of reason in the Anthropocene.

Thesis 69: Finally, the relational process of speculative reason links the world as a perspectival hyperimage in motion from which nature is conceived as multinaturalist circumstance from its immediately relational multiplicity.

Thesis 70: The perspectivism of speculative reason, in its multinaturalist link with the world, enables the capacity of simultaneous perception according to incompatible, paradoxical and irreconcilable perspectives.

Thesis 71: Speculative reason is the relational, anthropophagic and autophagic perspective of shamanic materialism.

Thesis 72: Shamanic materialism links ancestrality, extinction and trance as speculative and multinaturalist perspectives of the dialectical hyperimage of movement.

Thesis 73: Incompatible, paradoxical and irreconcilable perspectives constitute the relational correlate of the dialectical hyperimage of movement.

Thesis 74: What exists in multinature involves a turbulent mythical flow that continues to pulse and to extinguish; it is the powerful volcanic life of beings and things, a life that, through the cinematic and audiovisual process, expresses them not as differently perceived self-identical entities but as immediately relational multiplicities.

Thesis 75: The radical nature of the contingency asserted by materialism implies recovering the idea that destruction and annihilation can exist without any counterpart, without any compensation; it is about recognizing that there are absolute losses that will never be recovered, events without any consequences, civilizations and entire peoples annihilated and lost in the nothingness of History without leaving any trace, like those great rivers that disappear in the sands of a planetary desert. The idea that in the world there are a number of things, objects, subjects, events, flows, lives and events that will be absolutely annihilated and not recovered, the reality of that idea, we say, manifests the abrasiveness of the vast machinery of annihilation that inhabits the desert of the real.

Thesis 76: The perspectivism, the conversion and the mythical connection of the dialectical hyperimage open an eventual field in History.

Thesis 77: Shamanic Materialism as a cinematic and audiovisual process leads to and entails a political film-trance of agitation in which the agitation would no longer emanate from a coming to awareness or

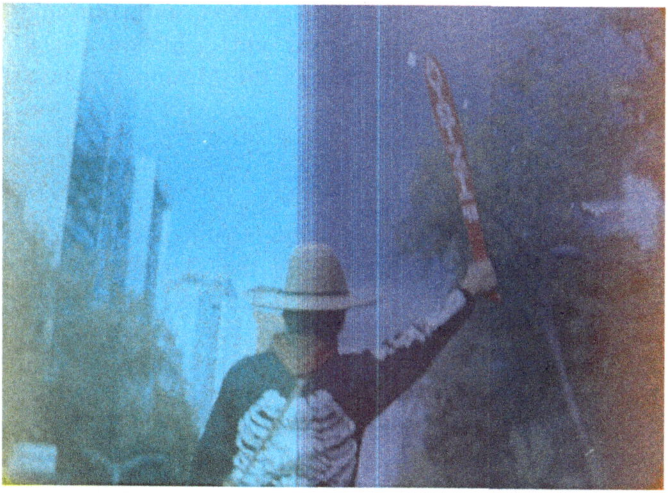

Fig. 16.1 Colectivo Los Ingrávidos, *The Sun Quartet, Part 2: San Juan River* (2017) © All rights reserved by the artist/Courtesy of Light Cone

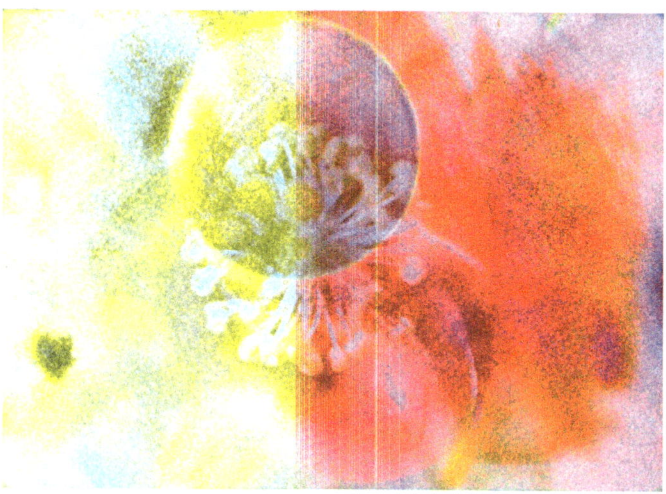

Fig. 16.2 Colectivo Los Ingrávidos, *Tonalli* (2021) © All rights reserved by the artist/Courtesy of Light Cone

call for mass mobilization, but would consist of placing contingencies and circumstances in a trance, including, in the first place, Reason itself, relating violence and contingencies to each other, so that through trance, violence, myth, ancestry, precariousness and extinction a new cosmological and cosmopolitical perception of the Missing People is formed (Figs. 16.1 and 16.2).

Notes

1. The space of media representation that corporate governments have built around them excludes the People. What corporate governments presuppose as a correlate of their own unpopular image is a population, a mass that they can administer, quantify, manage. The consistency of this space of representation is completely homogeneous, overdetermined, quantified and non-antagonistic. The determining dimension of the official discourse aims to model a total, coherent image that ends once and for all replacing or transforming the Missing People and the Supposed People into a Population. What the image emporium of corporate governments presupposes is a spectator population or an expectant population, a mass of hallucinated people that responds out of the unison of a passive civility to the question: what is behind the governmental or corporate images and audiovisual productions? Or: to what expectant totality do the monstrous stagings such as those continually undertaken by world corporate governments and their tele-computerized allies against all possible construction and articulation of Missing Peoples refer? The quantified, subjugated mass intimidated by these questions answers without a voice and from the rostrum of its atomization, responds as a Population. Behind the sound images that corporate cinemas and government cinemas disseminate, behind the hegemonic empowerment of the corporate-sovereign mannequins, there is an ideal of a spectator: a regulated population without a clear political territory, in a desolate environment of insecurity, as well as in a vertiginous aesthetic-political uprooting.
2. The determination of these three shamanic procedures involves three specifications: *perspectivist* possession, *transformational* conversion and *speculative* bonding.

3. One of the non-impure and absolutive correlates of the Aesthetics of Trance is the funeral liturgy.
4. The materialistic part of shamanic materialism forces us to place all audiovisual materials in a horizon of radical contingency as a field of indiscernibility common to all the collective contents of the image; this mobilizes us to assume an autophagic principle of relationality between images and sounds that are absolutely contingent, which in turn enables us to unleash a rhythmic outburst dependent, on the one hand, on a radical non-commutativity in the order of a temporality fractured by iterative modules, and on the other, on a constant tension between the apparent impossibility of continuous commutation and the reality of an expansive recursion as a horizon of formal determinability of the collective contents of the image. This brings us to the shamanic aspect of shamanic materialism through which we can intervene, induce and connect a trance as a rhythmic and bodily rapture on the determinability of audiovisual material according to an eidetic, archaic, ancestral and mythical differentiality.
5. This unleashing deploys an intermittent and glowing interrogation that spans centuries through time, reaching us in an extremely current moment of danger and perplexity; the aforementioned Mesoamerican spell prompts us to ask: *What land is this?* The origin of the question lies in the astonishment that the mercenary and later chronicler Bernal Díaz del Castillo expressed when entering an "unknown" land. In the heart of an "unknown" and "unnamed" America, the European Díaz del Castillo made clear the crisis of his colonial somnambulism, fusing his astonished perception with the intermittent glow of a non-European reality; here what the Europeans really discovered was the perplexity of their own ignorance. His ecstatic question crossed the destructive centuries of colonialism and European barbarism to be taken up, among others, by the writer Juan Rulfo, who recovered it from abandonment in the wasteland and the devastated lands left behind by five centuries of civilizing excess. Now we can recover the question, not only from perplexity and wasteland but from trance and cinematic rapture. Thus, from an intermittent and glittering land, we also have the trance as "images of relation". We may recall that the "letters of relation" were the way that Hernan Cortés informed the Spanish Crown about the "progress" of his ominous

conquest. Let us therefore recover the form of the "relation" as a report intended to *in-form*, but as the *un-formed* found in multiple "images of relation". Thus, through audiovisuals we seek to think, articulate, integrate and connect the historical absence of reconciliation entailed by the arrival of Europeans to the land later known as America as well as the destruction of Tenochtitlan in 1521. Five centuries after these events, we assert the Aesthetics of Trance and Shamanic Materialism as forms of multinaturalist resistance that move us to re-appropriate, transfer and transform the ominous corrosive process induced by that historical and asymmetric clash between radically different worldviews and which brought about the birth and development of modern nation states in the Americas with their resulting violence and problems that have still not been reconciled today.

6. *The circumstance* is the point of radical connection between contingency and facticity from which reason emerges. It is this situational sense of the circumstance in which reason acquires a possible determinability at the same time that it appears as a larval dimension that accompanies every entity, process and becoming (it should also be noted that the situational sense of the circumstance determines *place, position and perspective*). This perspective of nesting and infestation of reason supposes the nihilization of something given. Reason, therefore, is, in principle, a lack of being with respect to a given being and not the emergence of a full being; this parasitic dimension of reason constitutes its expansive infra-being, that nothing of being, nothing of History in which contingency and facticity are *linked*.
7. *Multi-nature* is a field of shamanic intersection between a constant epistemology and variable ontologies.
8. Currently, cinema lies between the zombie fetishism of the film, the instrumental tourism of museum finiteness, the tyranny of digital high definition of the industry, the simulation and forensic immersion of reality and the viral realm of low definition. By failing to take into account the eidetic chiasm produced by the arrival of Television for Cinema, the Internet for Television and the Pandemic for the Internet, Cinema is shipwrecked between a unilateral ideological drift and a neoliberal disengagement.
9. The cognitive and ontological catastrophe that neoliberal capitalism supposes and implements subtracts from the nature of

experience, thus revealing the objectifying emptiness of being, while in the audiovisual this subtraction leads to the double form of aberration: the sinister and estrangement.

10. Against the grain of *cognitive fracking* as a productivist activity of corporate control, we would like to point out and assert a series of audiovisual strategies (which are continually in danger of being exploited by neoliberal fracking itself) that come from three filmic (un)forms of industrial cinema. (1) *The Blob* (1988) by Chuck Russell: absorption, assimilation and expansion. (2) *Alien* (1979) by Ridley Scott: accumulation, nesting and infestation. (3) *Predator* (1987) by John McTiernan: strategy, camouflage and ritual.
11. The retroactive affirmation of planetary geological and atmospheric effects as clear causes of the corrosive effectiveness of the Anthropocene-Capitalocene determines the expansive scope of the linking and relational activity of Reason as an objectifying activity, as a faculty of the unconditioned, as expansion, generation and linking of categorical fields in the direction of the conformation of object inlays, eidetic retractions, intentional permutations, as well as cognitive catastrophes and ecological extinctions. All of this is articulated by Reason in the midst of a clear and continual contraction of hydrocarbons and energy that is increasingly radical.
12. Precariousness implies a qualitative and eidetic retraction of objects, which enables an intermediate and connecting constructivity between relational qualities that objects emit and absorb and that allows us to experience the shamanic constructivism of the audiovisual process from a synthetic point of view as additive and subtractive facets of qualitative and objectual incrustations, clearing a whole field of audiovisual entities and cinematic appropriations (dis)oriented towards objects.
13. This objectifying perspective also supposes and considers work as an anthropological invariant as a practical principle of transformation and effective production of the world. Would work as an anthropological invariant not suppose, in turn and as a correlate of its apparent invariance, a speculative invariant (perhaps more metaphysical than speculative) as a horizon of telematic determination present in multiple layers of human and posthuman history? Wouldn't Reason, as a faculty of the unconditioned, be precisely the aberrant retroactivity of a radical affirmation that determines the conditions of real experience? It is no coincidence that, in

the anti-ecological model of modernity, the pyramid is the figure that determines the representation of the food chain, symptomatically establishing a pharaonic condition of transcendent illusion for the current civil disassociation of the Capitalocene, thus producing not only pharaonic resurrection machines (corporate pyramids) but the objectifying assemblages of the sinister machine of capitalist precarization, devastation and annihilation.

CHAPTER 17

In Plain Sight, Out of Reach: Karrabing Films and Totemic Obligations

Elizabeth A. Povinelli

I

Years ago, I was at a weeklong conference on philosophical and anthropological approaches to ontological turn and how it expanded academic understandings of human nature(s). I was in the midst of thinking about a new book, which would eventually come out as *Geontologies: A Requiem to Late Liberalism*,[1] focused on the concept of geontopower. I was also finishing the final edits of the first film of the Karrabing Film Collective, *When the Dogs Talked* (2014). The film follows an extended family as they travel from their overcrowded government flat in Darwin, Australia to their ancestral lands along the coast hundreds of kilometres away. If they cannot find and bring back their relative, who holds the lease on the flat and has gone missing, they will be evicted and join the large numbers of "long grass mob" (Indigenous homeless who live on the fringes of

E. A. Povinelli (✉)
Columbia University, New York, NY, USA
e-mail: ep2122@columbia.edu

E. Della Noce and L. Murari (eds.), *Expanded Nature*, Experimental Film and Artists' Moving Image,
https://doi.org/10.1007/978-3-031-70729-2_17

Darwin). When the group makes it to their outstation camp, their relatives there divert them from their mission. Instead of saving their city flat, they are persuaded to go further into the bush to map the ancestral travels of ***debindurlg*** (usually translated as Wild Dog Dreaming or Wild Dog Totem) as part of an educational exercise for their children.

In a break between debates about the disciplinary ownership and political relevance of ontology during the day, some of the conference attendees asked if I would screen the rough edit. Immediately afterwards, and then over the next couple of days, a small discussion ensued about whether *When the Dogs Talked* exemplified the travails multiperspectivism approaches to nature face as they travel across global liberalism and capitalism, or whether it staged the fundamental differences among naturalist, analogist, totemic, and animist understandings of nature. Side conversations also included whether the film meant to intervene in debates about more-than-human forms of thought and intentionality and how, no matter its intention, the film spoke to debates about bio- and geosemiotics.

The side conversations about the relevance of *When the Dogs Talked* to philosophical and anthropological approaches to nature and ontology were interesting to me, not merely for clarifying the circuits that connected Paris to São Paulo, creating a rich debate between Philippe Descola, Eduardo Viveiros de Castro, and Eduardo Kohn. They also raised the question of what Karrabing intended to do with their arts and filmmaking. Were our films interested in entering academic or artistic debates? How were they responding, in content, form, and topic, to the late liberalism's governance of difference, including how this form of governance informed academic and artistic research and reception?

2

We have now made eight films, four or five short statement videos, and numerous installation pieces built around these films and videos. For all their variety in topic and genre, all Karrabing films are constantly reflecting on the nature and manner of thieves and thievery in late settler liberalism. Several films have theft as their explicit thematic. *Windjarrameru, the Stealing C*nt$* (2015), for instance, follows a group of four young Indigenous men who find two cartons of beer in the bush only to become entangled in a tense standoff between the police and their relatives (Fig. 17.1).

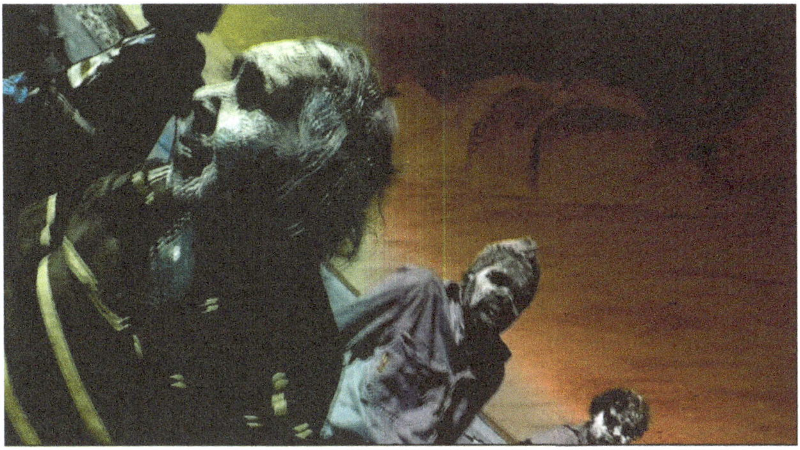

Fig. 17.1 Karrabing Film Collective, *Windjarrameru, the Stealing C*nt$* (2015) © All rights reserved by the artist/Courtesy of Light Cone

At stake is who is stealing what—did the young men steal the beer, which they insist they did not, or did miners blasting near a sacred site "set a trap" for the men to divert attention from their illegal activities? If you are Karrabing, you already know the answer to this question before the opening scene ends. In the Northern Territory, certainly to Indigenous communities throughout Western Australia, the meta-thematic of *Windjarrameru* is obvious—who goes to jail for what kind of thieving—as extractive capitalism wields enormous economic power. Two years before our filming of *Windjarrameru*, the Aboriginal Areas Protection Authority (AAPA) had sued OM Holdings for desecrating a sacred site on Kunapa lands. The AAPA acknowledged that the mining company's excavations did not overstep the literal boundary of the sacred site. Instead it argued that the company knowingly cut so deep and closely down the registered boundary of the site that its geological support was undermined, causing Two Women Sitting Down to crack, part of them falling into the manganese pit. Around the same time, the Indigenous Borroloola community was publicizing the contamination of their groundwater and rivers by the MacArthur River Mine, part of Glencore, a multinational mining corporation.

Other Karrabing films make reference to the constant pressure the state and capital puts on Indigenous people to open their lands to mineral extraction and the toxic results that follow. For instance, in *The Jealous One* (2017), the protagonist Rex Edmunds, on the way to a mortuary ceremony on his land, cautions his nephew never to listen to white people. "They talk a lot of shit, you know. They say this and that about your land. They sweeten you up with money. Then they tell you they want to mine on your land." In *Day in the Life* (2020), the same Rex Edmunds raps about having reformed his ways. "I used to drink. I used to smoke. I was that kind of bloke… But I've settled these days. I teach the kids the blackfella ways." The film shows a montage of Rex hunting and fishing before settling on him walking in the bush with his nephew, Gavin Bianamu. They size up tree after tree in search of the right one to carve a turtle and dugon harpoon, only to stumble upon a wildcat lithium mine. With a nice off-rhyme, Rex changes his tune. "How am I supposed to teach my kids, when white people digging on my land? No matter how much I try, this land is gonna die." In order to combat the assault on his land, Rex takes Gavin to a place where his male ancestors had regularly gathered to drink and practice their traditional songs. Rex calls out to his ancestors. He knows they are standing beside him although he cannot see them. (Film viewers see these ancestors montaged into nearby bushes.) One young ancestral man, moved by Rex's pleas, convinces his ancestral family to help Rex by singing a traditional song that will summon a great wind and storm, blowing white people out of Indigenous lands.

If *Day in the Life* focuses on material theft—the theft of Indigenous children, land, and capital—it is subtended but another mode of thievery, rarely explicitly thematized but always being staged in our films. This mode of theft has a deep history across the colonized world. It is the appropriation knowledge that canalizes and circulates epistemological value to the end of building disciplines. This disciplinary building can leak into public debates and attitudes that harden into state demands. They can then return to the same Indigenous communities as a toxic harvest. One of the most infamous and consequential instances of this began in the late nineteenth century when Francis James Gillen, the telegraph stationmaster at Alice Springs, Australia, and Walter Baldwin Spencer, the Chair of Biology at the University of Melbourne, offered to feed the Arrernte, Anmatyerr, Kaytetye, Warumungu, Luritja, and Arabana people, and protect them from settlers and police as long as they performed their ceremonial practices. Extracted from deep within the savagery of settler expansion, these forms of Indigenous obligation to their more-than-human relations were detached as information formed and reformed for the purpose of disciplinary building in sociology, anthropology, and psychoanalysis.

Karrabing don't live in a settler world structured by such crude forms of epistemological extraction and abstraction. No settler is usually poised at the fence, poison in one hand and a gun in the other, demanding Indigenous people perform as the price for their life. Nevertheless, Karrabing understand that their films sit within the long arm of this economy of culture and state of settler demands. The Karrabing Film Collective began in a period defined by a crudely racist assault on Indigenous "traditional culture". On the back of sex panic about unverified, unquantified, later discounted innuendos about Indigenous men's abuse of Indigenous children, the Australian federal government passed the Northern Territory Emergency Response Act, 2007. The Act didn't address sexual abuse. Instead, it seized control of Indigenous communities and lands; flooded communities with police; banned pornography; removed customary law consideration from bail and sentencing applications; removed the permit system whereby Indigenous people controlled who could or could not enter their lands; put strict limits on what welfare payments could be spent; and abolished federal financed work programmes. The aim was to push Indigenous people into low-paying jobs and open their lands to mining as it created new consulting and police jobs for a mostly white employment force. It was declared the great upending of the era of Indigenous self-determination.

This supposed upending of Indigenous self-determination inserted its architecture into a much older form of settler recognition. The Indigenous owners being pressured to open their lands to market-based extractions were persons who the state had determined to be the "traditional Aboriginal owners". "Traditional Aboriginal owner" is a legal term which has become a shorthand ("TOs") for an Indigenous form of land ownership written into the Aboriginal Land Rights (Northern Territory) Act,[2] 1976, based on a social anthropological model of the totem-clan-land ("a local descent group of Aboriginals who have common spiritual affiliations to a site on the land, being affiliations that place the group under a primary spiritual responsibility for that site and the land") that nicely fit liberal understandings of sovereignty and property. It also severed the connections between people and their ancestral relations based on totemic travels, bodily and ecological interdependencies, and ties of marriage and ceremony. As the same Rex Edmunds noted in *The Riot* (2017), the Land Rights Act allowed everyone to have "a little bit of land" as it sliced through mutual obligations that kept each land in its form. Animals, sands, subsoils, water flows above and below ground stay in place by remaining connected through other places just as ancestral totems created the shape of places because of events that happened there and at other places. And since the law made knowledge about place the currency for recognition, it incentivized both stealing and withholding knowledge—pitting Indigenous people against each other. Karrabing are no strangers to this. Even as we were beginning filming *The Jealous One*, the anthropological wing of a land council tried to circumvent explicit instructions from members that they not seek access to an important regional narrative from another Indigenous bureaucracy. Karrabing knew on past experience that, once in the land council archive, few would respect the "classified" stamp. This is the real dirty politics of knowledge, to which Edmunds is referring, when money is at stake.

All Karrabing films must grapple with a specific problem about how to orient its land and artistic practices within and against this unrelenting history of knowledge extraction, abstraction, and circulation that moves value out of and away from its purpose, namely, the strengthening of human obligations to the ancestral present. In this sense, some of the comments about *When the Dogs Talked* at the philosophical and anthropological colloquium were quite insightful. The making of the film is a reiteration of the content of the film—as we tell a story about a group being diverted from the path of settler property to the path an ancestral group of dogs made, Karrabing is enacting this diversion. In making the film, they are removing themselves from the relentless demands of the settler state—the topic of the film—and conforming their lives, instead, to the path laid down by the Ancestral Dogs.

3

As Karrabing films began to follow another path—the infrastructural paths of film festivals, galleries, and museums—a different aspect of theft arose, namely, that of cultural appropriation. Would Karrabing give in to the soft and hard desires of cultural institutions and their clientele to act as a translation machine—a soft version of Spencer and Gillen's devilish deal with the Arrernte and other Central Desert peoples? If not, how could we make films that enacted and transmitted Karrabing's understanding of the mutual entailment of their human and more-than-human worlds, thereby deepening their ancestral obligations, without revealing this understanding to anyone who could afford a ticket to a screening? In response to this problem, the Collective has tried to create a filmic form in which a film or installation shows the ancestral present of Karrabing worlds without providing all of the crucial interpretative frameworks for understanding what is being shown. Numerous decolonizing curators allied with our effort. This strategy is referred to in our fourth film, *The Jealous One*, commissioned in part by Vivian Ziherl's Frontier Imaginaries curatorial project. A chapter title toward the end of the film states, "if you know…where they all meet…why, and what always happens". By showing without giving it all away, Karrabing films press against settler demands for them to be transparent to settler appropriation and strengthen their hermeneutic control. Two examples help illuminate this dynamic.

The first example comes from our first screenings of *When the Dogs Talked* at Gertrude Contemporary in Melbourne. Seven of us gathered there were asked, given the calamities that befall the human protagonists, whether the Dog spirits were good or bad. When Cecilia Lewis, puzzled by the question, asked for clarification, good and bad shifted to benevolent or malevolent. Cecilia answered that they are neither good nor bad. They are Dogs. No one in Karrabing with whom I have talked about this moment thinks the questioner was somehow bad to have asked. "They just don't know how to think," is the common response. When we move outside Australia this inability to find a frame through which they can make sense of what they are hearing and seeing can be even more difficult, especially when we refer to ancestors and Dreamings. Spirituality emerges as a translational corridor leading to abstracted entities experienced in dreamlike states haunting Karrabing realities. Karrabing members have responded to these confusions in various ways. For instance, a collective agreement has emerged in the wake of such interpretive guesses about the meaning of "the Dreaming", to use the term totem. As Gavin Bianamu said to me in New York, "Just say totem. They can get their brains around that a bit." His sister from another mother, Natasha Bigfoot Lewis, also patiently explains how her Barramundi totem off the coast of her land is related to a whirlpool dreaming down the coast to the east and to another Barramundi somewhere up a large estuarine creek. In other words, the general analytics of Karrabing existence is considered important for Indigenous and non-Indigenous people to understand if there is to be any hope intervening in the ongoing collapse of the human and more-than-human world under the strain of settler liberalism and capitalism.

At stake is not merely the material mutual obligation of ancestral landscapes but the affective logics that subtend existence. A second example comes from queries that sometimes arises after screenings of *Wutharr, Saltwater Dreams* (2016). *Wutharr* is based on a boat trip to Karrabing's remote coastal lands. When the motor broke, five of us were stranded in one area, seven or eight in another. A high-end white fishing tour floated by, who by law are allowed to take fish from Karrabing waters as long as they never set foot on the land. When those of us stranded without fresh water nearby asked them for some, they told us to save ourselves. Other funny and not so funny trials and tribulations followed before the motor decided to work again and we all made our way back across the bay to where we were camping. During the long motor back, we debated why the boat had stopped running—the ancestors were punishing us for not visiting; Jesus was testing our faith; we couldn't properly service the boat given scarce economic resources (Fig. 17.2).

When we screen *Wutharr*, we often ask questions about specific moments in the film when a chorus of ancestors tracking the protagonists chant "punish them, punish them" because the protagonists have

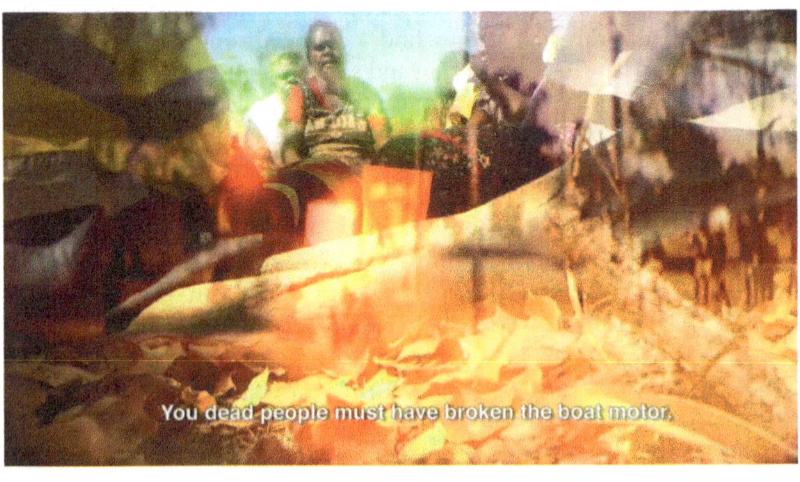

Fig. 17.2 Karrabing Film Collective, *Wutharr, Saltwater Dreams* (2016) © All rights reserved by the artist/Courtesy of Light Cone

not visited them for a long time—and proceed to do so by hiding water holes and breaking motor wires. These are hysterically funny parts of the film. And the high-pitched chants are hard to get out of one's mind after hearing them over and over. The interpretation some viewers take away is that if our erstwhile protagonists had done the right thing, then they wouldn't have been punished. We have explained numerous, sometimes seemingly endless, times that, no, ancestors like all other aspects of existence are jealous, and that jealousy is a ticking time bomb that starts the moment one turns away from anything or anyone and thus something one must deal with rather than believe can be dealt with. But as we have done so we have also come to understand not merely the difference between philosophies of desire and Karrabing analytics of jealousy but the stakes of the difference in the practice of obligation. You might wonder what the stakes of these differences are. If you know...then you know...

This is, of course, misleading. Knowledge is not, however, something you know or don't know. Knowledge is—or Karrabing attempt to keep knowledge as something that is—earned through a life of orientation to an ancestral present, earned within specific ways of listening and looking, piecing together, and testing veracity. Karrabing like what Vine Deloria Jr said about the difference between white and Native forms of revelation, the one built on a model of truth as an event that changes the time and space of all existence, the other on a model of truth as an understanding of a shift in modes of relation and thus a call for the adjustment in how we are reciprocally obligated in these new relations.[3] If, Deloria argues, Western forms of revelation promise a universal truth applicable to each and every individual, Native revelation seeks to understand how a previous understanding of the relations between elements of existence have altered and how this alteration demands a new understanding of how local worlds are held together or at risk of breaking apart. Native revelation does not seek to abstract and then impose on all others their ways of relating to each other and the more-than-human world. Instead, they place these local relations in relation to other local relations. We can think here of Aimé Césaire's reflections on the function of the universal in colonialism. "The universal", he said, "is not the negation of the particular but is reached by a deeper exploration of the particular".[4] He continued, "The West told us that in order to be universal we had to start by denying that we were Black. I, on the contrary, said to myself that the more we were Black, the more universal we would be."[5]

At stake is a question of how to convey specific stories about specific ancestral presents for specific relations while supporting this kind of relation-oriented knowledge. At the core of all Karrabing films is thus a multifocal addressivity that assumes not everyone knows what exactly is going between the levels of human and ancestral relation but they don't get to and they can't because, as a theorist once said, the map is not the territory. How can you understand how the sounds relate to the characters if you cannot distinguish a crow from a chicken hawk, a pelican from a baby's cry, a fly and a bee? In order to understand the characters and plots of Karrabing films, a viewer needs to have an intimate knowledge of the semiotics of local flora and fauna as well as local metapragmatic orders that link these ecological semiotics to totemic and human geographies. Without these two orders of knowledge and an attentiveness to how each might be altering in relation to the other, a viewer can be shown an entirety of Karrabing existence without understanding what they are seeing or hearing.

Most importantly, what relations to humans and the more-than-human world provided you with these ways of seeing, hearing and moving, and how are you fulfilling your obligation of keeping them in place as you find your place?

Notes

1. Elizabeth A. Povinelli, *Geontologies: A Requiem to Late Liberalism*, Durham, Duke University Press, 2016.
2. "A group of local Aboriginal ancestry which has common spiritual affiliations with a site on a given piece of land, being affiliations which place the group under a spiritual responsibility for that site and for the land", The Aboriginal Land Rights (Northern Territory) Act, 1976 (ALRA).
3. See Vine Deloria Jr., *God is Red: A Native View of Religion*, Wheat Ridge, Fulcrum Publishing, 2016. And Zoe Todd, "Fish, Kin, Hope: Tending to Water Violations in amiskwaciwaskahikan and Treaty Six Territory", *Afterall, A Journal of Art, Context and Enquiry*, no. 43, 2017, pp. 102–107.

4. Annick Thebia Melsan, "The Liberating Power of Words: An Interview with Poet Aimé Césaire", *Journal of Pan African Studies* 2, no. 4, 2008, http://www.jpanafrican.org/docs/vol2no4/2.4_The_Liberating_Power_of_Words.pdf, p. 5.
5. *Ibid.*

CHAPTER 18

The Figure of the Peasant Farmer-Researcher

Gérard Leblanc

The history of the human species' relationship with nature is nothing other than the history of the forms of interaction it has constructed with it. Does this mean that the human species and nature are on the same level? Despite all our efforts to do so, are we capable of destroying nature, or rather our own conditions of existence in nature? The end of the world envisaged by many will only ever be, if it actually happens, the end of human presence in the universe. Even if our planet were to be abolished, the life of the galaxies would nonetheless continue its course.

Agriculture certainly exerts a form of violence on nature insofar as it organizes it for purposes that are not its own. Nothing is produced or consumed in nature that is not immediately recycled. Nature knows no waste. Hunter-gatherers only took from nature that which corresponded to their vital needs in what was spontaneously produced. They had no intention of transforming it. They acted in the same way as other animals,

G. Leblanc (✉)
IRCAV, University Paris 3, France
e-mail: gerardv.leblanc@gmail.com

Médias Création Recherche, Langres, France

E. Della Noce and L. Murari (eds.), *Expanded Nature*, Experimental Film and Artists' Moving Image,
https://doi.org/10.1007/978-3-031-70729-2_18

with the reversible roles of predator and prey. Nevertheless, the cerebral superiority of the human species, evident in the manufacture of weapons and tools, quickly materialized in its desire to eliminate more physically powerful animal species. It was no longer just a matter of fighting predatory individuals, but of exterminating the species from which they originated. Very early on, the human species wanted to dominate other animal and plant species.

Hunter-gatherers had no knowledge of agriculture. As nomads, they moved to areas where the conditions for their nutrition were the most favourable. This is what all animals do. Agriculture was born with sedentarization. It began as subsistence farming. You eat where you live and you consume everything you produce. You do not produce beyond the needs of small communities. The cultivation of land, often reclaimed from the forest, had little impact on natural ecosystems. The planet was sparsely populated and the means of production rudimentary. Of course, chemical products did not yet exist.

Slavery and then feudal social relations remained based on subsistence agriculture. However, the producers no longer produced for themselves but for the ruling classes. They found it extremely difficult to survive off their work. The negative impact on natural ecosystems began to be felt, as nature was treated with as little consideration as the producers.

The idea of human domination over nature was not born with capitalism. But it has industrialized it. It is now a matter of producing for national and international markets. A dominated country can produce for export without satisfying its own needs for subsistence crops. Agriculture has become an industry that must demonstrate its competitiveness on a global scale. To bring down the price of agricultural goods, yields must be increased. This was achieved through the mechanization of ever larger plots of land and the increasingly intensive use of chemicals, starting with ammonium nitrate, first tested as a weapon of mass destruction by the German army during the First World War. The earth is considered as a base from which all biological and organic life presumed hostile to the increasingly artificial crops being planted must be eradicated. This led to the death of soils, and yields, which had been initially boosted, ended up stagnating and then regressing.

To radically change our point of view. To take nature's point of view, to listen to it. Sight and hearing. What does this mean?

Nature speaks through the peasant farmers who interact with it. When working the earth, listening to nature means working with nature. It means realizing that nature is the site of an infinite multitude of transformative processes. It means trying to feel and understand what it is doing. It means seeing things from its point of view. It means ensuring that nature finds equivalents in human language that can shed light on its practices.

What is nature telling us that we do not hear? And first of all, what are we asking of it? What does it mean to listen to nature? It is in the practice of peasant farmer-researchers that we find elements of an answer.

Un autre horizon (2010), *L'autonomie paysanne* (2013), *L'horizon des possibles* (2016). These three films, co-directed with Catherine Guéneau, aim to sketch out the figure of the peasant farmer-researcher who asserts themself through organic farming, in opposition to the discourses and practices imposed by the agri-food industry and chemical technicians. The aim is henceforth to work with nature and not against it. This means returning to nature to try to understand its laws. It means studying the models it offers us, models that have lasted for so many millions of years, such as the forest.

Start from the questions posed to research by the producers themselves. Questions to which they provide answers in the field (hence the figure of the peasant farmer-researcher) in conjunction with professional researchers, often disconnected from the institution rupture (a disconnection which they did not seek but which was imposed on them by the labelled orientations of research). We wanted to show that research starts in the field and then returns to it. There is no separation. Production and experimentation interact.

Our problem as filmmakers was to find a *form* that could materialize this unity between the field and research. We therefore had to start from concrete situations (any filmed speech would problematize questions encountered in the field) and eliminating any overbearing or connective voiceover. We wanted to be able to move from one type of crop to another, making it possible to appreciate and understand that the problems that arise in terms of soil fertilization and crop protection are both different and identical. The films would bear witness to encounters that were already taking place and, through their editing, would create new ones.

Peasant farmer-researchers have things to say. These things have been felt, matured, turned over and over in the minds of the peasant farmers with whom we have set out to work. The result, for those who speak, is a discourse that is assured of its foundations as the architecture of a well-designed and well-constructed building. It is a discourse that has matured like a fruit that can be savoured *right here and now*. And it is exactly here that this discourse was cultivated: in this field of triticale or in this field of rye. Of course, it did not spontaneously grow on its own or from its immediate environment, it was also nourished by external inputs, sometimes from far away. But these inputs have been assimilated and interpreted according to the specific cultivation conditions of each farm. What works for one farm does not necessarily work for another, even if they are only a few kilometres apart. The soil can be very different from one plot to another, sometimes on the same farm.

We quickly realized that peasant farmer-researchers do not use ready-made thoughts or ready-made plants. They do not apply recipes. They know that nature is complex, and they do not try to reduce this complexity by applying the chemical yoke that is supposed to take care of everything. They do not seek to tame supposedly dangerous nature, but rather seek to accompany it by gaining a better understanding of the contradictions that cause it to develop outside of all Manichean frameworks.

There is nothing dogmatic in the peasant farmers' discourse. Their statements affirming orientations and positions, spoken with assurance, are based on research and working hypotheses that always need to be verified in the field. There are often possibilities, sometimes probabilities, never certainties. This is not to say that their discourse is unsure of its directions and positions: it is these that lead it to construct hypotheses.

A well-grounded discourse is not just about meaning and concepts. It is also percepts and affects. It is this mixture, as we say for intercropping, that gives the discourse its emotional charge and makes it immediately shareable and appropriable. Here, emotions are not used to manipulate ideas, nor ideas to manipulate emotions. Emotions and ideas have grown together. They are continually intertwined and refined by contact with each other.

Our job as filmmakers? To elicit this discourse, to free it, to capture it in its emergence. To cross it with other discourses, captured under the same conditions. Cross discourses? Not just that: to make them interact. They reinforce, contradict and complement each other. To organize, through editing, encounters that have not yet taken place in reality. This, too, is the purpose of editing: to transform virtual encounters into real ones. To anticipate reality.

But there's another point worth remembering and exploring: films anchor discourse to the place from where it is emitted. This is a twofold place: there is the body that emits the discourse, but there is also the environment in which it is deployed. This environment is larger than the territory occupied by the farm. It is inscribed in a physical and mental landscape that it helps to reshape.

Both can be captured within the same frames. Our framing affirms the primacy of nature without erasing man, on the—essential—condition that he does not seek to dominate it. Simply to feel it, to understand it. But this *simply* is as vast as the research, always ongoing, never completed.

Our approach also aims to capture the sensibility of the relationship between our "actors" (actors of their own lives before being actors in our films) and their environment. This sensitive relationship manifests itself not only in the discourse but also in the gestures that accompany it.

Take, for example, the cereal farmer Lionel Caudy in his rye field during a tour of his plain where he confronts his working hypotheses with the questions and experiments of other peasant farmers (*L'autonomie paysanne*). As he talks, he begins to caress the ears of rye. The gentleness of the caress echoes the rigour of a discourse replete with questions. Without this love of nature, which materializes in the gentleness of the gesture, there would be no thought capable of being attuned to it.

In our films, it is the editing that acts as a voiceover. There is no voice from on high telling the viewer what they should hear and see in what they are told and shown. We are not opposed to the use of voiceovers in general, we oppose voiceovers that do the work for the viewer or that make the viewer's job easier. Refusing this type of voice requires editing that allows the viewers to think and speak for themselves, using the audio and visual cues provided by the films.

More and more of us believe that the development of organic farming will determine the future of humanity. In this sense, it is of course a vital issue.

There are fewer of us who think—and this is also what we wanted to make tangible and intelligible in and through the films—that the development of organic farming also gives rise to the deployment of a form of thought that resonates with nature and corresponds to its real movement: dialectical thought, derived from the dialectics of nature.

If, as soil microbiology researchers Lydia and Claude Bourguignon say in our film *Un autre horizon*, the forest is a natural model for agriculture, it is also a natural model for thought. For millions of years, and without any human intervention, it has continually resolved the contradictions that conditioned its development. We must adapt the natural model of the forest to the artificiality of the agricultural and viticultural models developed to meet man's food and taste needs, which are less often recognized in class societies where the reproduction of the material conditions of existence has always taken precedence over the pleasure of living and, consequently, among other things, of tasting.

We chose to conclude the series of discourses that question and respond to each other throughout *Un autre horizon* with this analysis by Lydia Bourguignon: "Man only uses nature to an infinitesimal percentage of its potential". We are a long way from knowing everything about it. Instead of acting as if we had nothing more to learn from it and can therefore replace it (genetically modified organisms are the target here), we could and should, if not driven by immediate profit interests, experiment with new crossbreeding and hybridizations between plants and animals. As Eric Petiot stresses in the same film, "Man should not replace nature". We must strive to find ways, also taken from nature, of stimulating plants' immune defences and promoting their growth.

We are far from knowing everything about nature, but that does not mean we know nothing about it. It is by working with nature that we learn to know it better and better. We sometimes observe phenomena that we cannot explain. While in the field once again, the winemaker Alain Réaut observes that the biodynamic methods he uses produce better quality compost, but for the moment he is unable to explain why, though he continues to try to understand the reasons. Thus research progresses, from empirical observations to hypotheses to be verified.

"To be a researcher, you have to be curious", notes the market gardener Xavier Deleau. This curiosity is shared by all the actors in our films. Jean-Pierre Fleury, one of the pioneers of biodynamic viticulture in the Champagne region, highlights the conjunction between his interest in astronomy and his passion for biodynamic viticulture. There is no need to descend into abstruse mysticism to recognize the influence of the stars on crop development.

Cereal farmer Jean-Pierre Cathelat has no shortage of curiosity either (Fig. 18.1).

Here he is in the middle of his wheat field. Today is a big day because he has just made a discovery. The thistles, which "we have to accept grow above the wheat", are dying without him having to use the slightest chemical product to eradicate them. The discovery? "Thistles are our

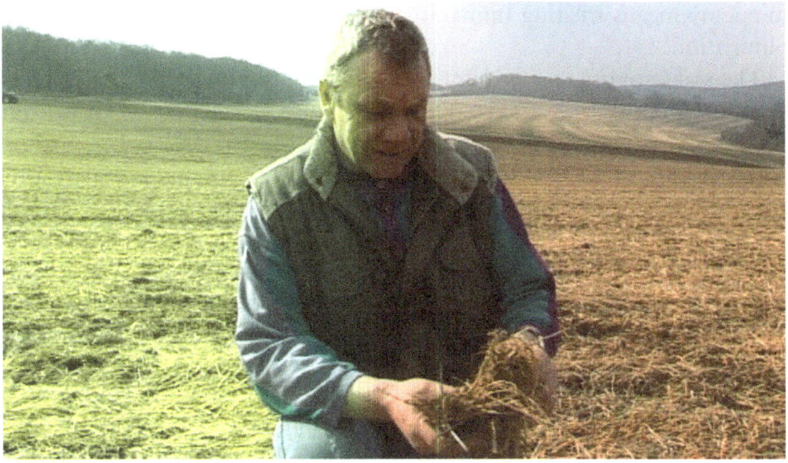

Fig. 18.1 Gérard Leblanc and Catherine Guéneau, *Un autre horizon* (2010) © All rights reserved by the artist/Courtesy of Light Cone

enemies, but they have enemies too". In this case, a caterpillar that only attacks thistles and ignores the wheat. Observing nature leads to discovering its dialectics and to thinking dialectically.

Organic farming and dialectical thinking are therefore inextricably linked. They stand in opposition to all the Manichean approaches that stem from the interests of the chemical agri-food industry, which would have nature be the enemy of the human species.

In a 1978 film (*Avenirs*, directed by Ada Rémy), the now defunct chemical company Rhône-Poulenc had relied on such Manichean approaches to better sell its fungicides, pesticides and other insecticides. There was no question, for example, of dividing insects into "friends" and "enemies", also capable of fighting each other, either directly or indirectly through the food chain. They all had to be eliminated. Only, here's the thing: they were becoming increasingly resistant to the chemical products that were supposed to be able to destroy them. So what was to be done? The answer lay in Rhône-Poulenc's short-term interests: ever more powerful products had to be developed in the hope that one day, after however many generations of products, no insect would survive. There was just one thing they were forgetting: treating nature as an enemy means treating humanity as an enemy, ultimately leading to its elimination.

The stakes of organic farming are therefore also situated in the realm of thought, thought that aims to transform nature on the basis of knowledge of its laws.

In the prevailing mindset, organic farming is often associated with low productivity. Hence the—also prevalent—idea that we could not feed the planet with it. *L'horizon des possibles* dispels this nonsense. Using the example of market gardening, the film shows and demonstrates that it is possible to achieve much higher productivity than conventional so-called "intensive" market gardening. This is particularly true of permaculture. Why? Because you produce your own seeds, those that are best suited to the land you cultivate. Because you learn to enrich your soil with organic matter. Because you learn to assemble vegetables together in time and in space. Because you associate vegetable associations with other plants, flowers, shrubs, fruit and other trees and cereals. Because, as nature teaches, you create or recreate ecosystems based on biodiversity. If bodies of water are lacking in an ecosystem, you create them. They will produce organic matter, improve soil moisture, soften the rigours of the climate. The animals that live there (ducks, amphibians, insects) will play a useful role as predators. A balanced ecosystem is inconceivable without animals (Fig. 18.2).

It is certainly in permaculture that the figure of the peasant farmer-researcher culminates. It brings together two functions that class societies have separated and opposed: manual labour and intellectual labour. Manual labour plays a much larger role than in chemical agriculture. As the area worked by each peasant farmer is much smaller, there is little place for mechanization. There are few or no tractors (which, incidentally, reduces the space between sowing and helps improve productivity). No chemical products are used, to weed for example. Almost all operations are carried out by hand or with the help of farming implements adapted to the hand, often reshaped according to the needs of the soil, which are increasingly well identified. But this manual labour would be in vain if it were not informed and guided by intense intellectual work based on a sensitive relationship with nature and a continuous process of documentation and analysis. Permaculture is an aesthetic project that aims to make nature more *edible*. Charles-Hervé Gruyer, a permaculture market gardener, points out in *L'horizon des possibles* that productivity is an effect of a successful ecosystem that satisfies all the senses. The aesthetic

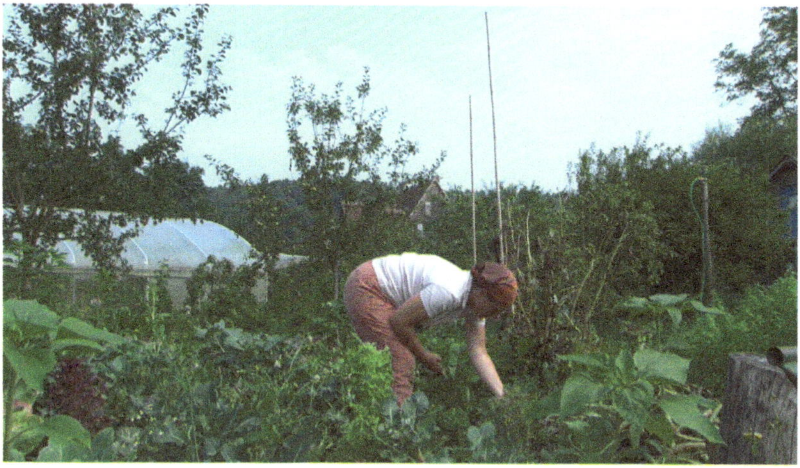

Fig. 18.2 Gérard Leblanc and Catherine Guéneau, *L'horizon des possibles* (2016)

sense encompasses all the others, including those of taste and smell. The concept of beauty is thereby renewed and strengthened. The visual and aural palette offered by nature, based on biodiversity and heterogeneity, is considerably broadened. Nature has never clamoured so much in its visual fullness.

All that remains is to film this nature transformed by peasant farmers who have agreed to let themselves be *cultivated* by it. If there is an aesthetic productivity of films, it will be provided in spades.

CHAPTER 19

For a Cinema *After Nature*: Jacques Perconte and the Art of Withdrawal

Frédéric Brayard

The 'ontological turn' has deconstructed the concept of nature, and the work of thinkers such as Philippe Descola,[1] Bruno Latour,[2] or Donna Haraway[3] suggests new paradigms in which we situate ourselves and our presence in a world 'after nature'. What about film? How to think a cinema after nature? How can film engage with the world without reproducing regimes of organization, and distributions of power, inherited from the systems of dual oppositions constructed around the concept of nature[4] (e.g., visible and invisible, seer and seen, subject and object, shot and reverse shot, absence and presence)? How could a possible 'cinema after nature' acknowledge the agency of all things, whatever they are, that signal themselves at different moments of the film's elaboration and give them the possibility to 'make film' as they "make world"?[5]

Answering such questions necessitates reloading the question of the ontology of cinema, to consider film as a producer of ontologies and to

F. Brayard (✉)
University of Central Lancashire, Preston, UK
e-mail: Fbrayard1@uclan.ac.uk

E. Della Noce and L. Murari (eds.), *Expanded Nature*, Experimental Film and Artists' Moving Image,
https://doi.org/10.1007/978-3-031-70729-2_19

analyze which mode(s) of being(s) it can recognize and which order(s) of things it can produce. In its classic version, (post)Bazinian film explores the complexity of possible correlations between its own presence in the world and the presence of the world for us.[6] However, digital cinema urges us to reconsider the situation since it needs neither present nor presence, and no longer directly produces images.[7] As Ángel Quintana explains, with the digital, "images have lost their negative. The original mould that preserved this imprint of the physical world has disappeared".[8] This indexical rupture frees the cinematic real from the supremacy of human modes of presence in the world[9] and allows the digital to develop a flat ontology[10] that considers all things equal from the perspective of their being and their agency. This opens the way to an ecocritical 'cinema after nature' as practised by Jacques Perconte.

His work resonates with a tradition of experimental cinema which, according to Nicole Brenez, "disturbs the received uses of images, invents or reconsiders their role, their plasticity, their speculative force …".[11] Such a cinema highlights various decisions, actants, and operations of power distribution that are central to but obscured by the production of its images. Perconte is part of a lineage of filmmakers who challenge the concept of the essence of nature, given and accessible in its a priori presence, available to the gaze of humans and cameras alike, by highlighting the labour involved in the filmic production of nature. Indeed, if nature exists, it is in excess of any image produced by a punctual, stabilizing, selective, and organizing subjective position. One cannot give accurate images of nature, but only images that give an account of our efforts to make nature by making films. This deconstruction of the '*im-media*-tely'[12] accessible presence of nature has been explored by a tradition of analogue filmmakers (Cécile Fontaine, Mahine Rouhi and Olivier Fouchard, Anne-Marie Cornu, and many others) through practices that reveal the presence and construction of images that make film; the frames can thus be dyed, painted, scratched, superimposed, filmed, etc. Conversely, digital moving images do not rely on the materiality of frames, and it is this disappearance of images that Perconte makes appear. By revealing, each in their own way, the unseen at work in the making of images of nature, both approaches share a similar questioning of the presumption that the presence of nature is directly accessible. This effort, however, is made, when working with analogue film, through a *reappropriation* of the materiality of the image and a focus on the ontological evidence of the trace. In this case, being is granted to whatever leaves an imprint on the frame at any moment of the process, but in which the human gesture (scratching, painting burying the frames, etc.) is, in the last instance, significant.

Perconte's approach, on the other hand, consists in *letting go*, abandoning the film to the world, and everything that makes sense to the insignificant digital signs. This is achieved through shooting practices that seek to create fields of equality that are opened to anything that can *be* and participate in the film making. Then, compression programmes translate the data collected during the shooting into space-times in which everything can create flows of colours, pixels, accelerate or freeze movement, and enter or not into interaction.[13] Such a cinema demonstrates that what is lost when acknowledging the disappearance of nature and of images is the centrality of human subjectivity. The digital allows us to shift from an ontology of the trace for us to a flat ontology that puts on the same level everything that is and can be, regardless of the classifications and distributions to which we can subject them (natural, cultural, true, false, (im)material, living, inert, etc.). Perconte's experimental approach is therefore twofold. It first consists in revealing the potential agency of actants we make world, but rarely make film with. It then highlights the processes of digital data translation that usually invisibilize these actants, and, by embracing alternative processes, allows them to participate in the film-making experiment. Perconte thus suggests a reality of which we are neither the centre, nor the condition that regulates its emergence. This reality is not the set of things that we consider as present to our presence in the world, nor the different traces that allow us to attest and evaluate these presences, but the possibility of being of any thing, and the possibility of all things to enter or not into relation.

Such a cinematographic practice therefore allows us to respond to the demands of a position after nature that Latour describes as follows:

> We have to agree to remain open to the dizzying otherness of existents, the list of which is not closed, and to multiply ways they have of existing or of relating among themselves, without regrouping them too quickly in some set, whatever it might be, and certainly not in "nature".[14]

We can then redefine nature and think, with William James that "[n]ature is but a name for excess",[15] the infinity of beings and of their ways of interacting and making worlds, the inexhaustible, that which escapes the correlation between the world and human consciousness, the unthinkable. It is possible to argue that a cinema that places the power of any actant, whatever it may be, at the heart of its relationship to the world is therefore 'after nature', both *posterior* to the binary regime of nature–culture and *according* to the contagious, incalculable, unthinkable creative force of the real.

In Perconte's specific practice, creating a filmic field of equality open to any actant begins with the shoot. Vincent Sorrel[16] explains how Perconte uses very long focal lengths not to reduce the field of the visible to what is far, but on the contrary, to account for distance and flatten it. The camera does not function like an eye that distinguishes better and which, by focusing, structures space, selects some objects and leaves others in blur; distance itself, and everything it includes, is included and flattened in the frame. In Perconte's words:

> [I] try to film distance itself ..., find ways of making sure that things can no longer be far away ... to bring everything to the surface of the screen. ... There is no more depth. I try to put the different elements in relation, to flatten the shots and to bring the elements closer so that one single space accommodates their luminous vibrations.[17]

The act of filming does not constitute one single seeing subject extracting themselves from a nature that they face. It is not about organizing space, identifying and revealing objects, nor establishing degrees of presence. It consists in cutting out a plane of equality in the space–time, flattening everything that it includes and considering as an actant anything that introduces change. According to Vincent Sorrel:

> The gesture consists in gathering what, in the details encountered in nature and in the events that occurred during recording, cooperates as life: dust, insects, optical aberrations, defects in perspective, currents and masses of air, but also accidents, technical unforeseen events.[18]

What is at stake is the the development of a set of practices that grants no-matter-what[19] an equal ontological status, and an equal agentivity in the film-making process. For instance, *Or / Or, Hawick* (2018)[20] mixes a detail of Klimt's *Kiss* and a sky in which a bird flies; while the camera follows this specific bird, the film is made with all the contingent events assembled by the shot: other birds that cross the frame and introduce variations in intensity in the generation of colour movement, insects that fly close to the lens and whose flight leave strokes of white pixels that winds and clouds slowly fill again with colours. *After the Fire* (2010)[21] reloads the phantom ride by filming a landscape from the very front of a moving train, in which spots or reflections on the locomotive window cause pixel flows to burst forth. Such practices acknowledge the equality of being and agency of everything, and open the film to all possible interactions, all links, transitions, that these actants can weave together. Since reality is no longer the sum of things that we touch by hand or eye, but the possibility of being and interacting of everything, making film with reality consists in celebrating the equality of everything that makes up the world, everything that the event of the shot can comprehend. The bird of *Or / Or, Hawick* (2018) is therefore not an object that a seeing subject follows in space and over time but an actant who, interacting with air masses and all the other actants that the film could neither list nor exhaust, directs the translation of the shot in space and the emergence of the events that occur there. In the fixed shots of *Forêt d'Allogny, Les Gaillards* (2016),[22] where trees with golden autumn leaves move in the wind, the distinction between objects and processes becomes inoperative, and again, everything (sky, leaves, branches, trunks, wind, animals, colours, camera, focal length, change of light, energy exchange, image compression) is equally a thing and equally makes film. The shot frees the possibility of being pure objects (a leaf, the sky), pure processes (the movement of branches in the wind that causes the leaves to detach), but also of all impurities since any object can become process and vice versa. Thus, in falling, the leaves can, in the course of their fall, either not affect the elements they encounter, or create blocks of grey pixels that persist and can then themselves follow the movement of the wind that moves the branches and mix their grey with adjacent things and colours. The space of the film is therefore relational; it is not an environment in which individualities persist but a variable space, transformed by the actants, whatever they may be, that signal themselves there. Nature is neither object nor subject, even less frame, not even environment. It becomes the set of changes, visible or invisible, continuous or discontinuous, that any actant can trigger in the ecosystem of the film. It is therefore indeed *after nature* that Perconte films, since, by setting up such a plane of equality, he creates a field of possibilities that reflects the way in which interactions between actants at work in the profilmic world can induce instabilities, evolutions,[23] insistences and resistances,[24] or extinctions.

In *The Movement-Image*, Gilles Deleuze facetiously lends to certain philosophers "war cries".[25] The cry that would concentrate Perconte's energy for making world when he makes film would not be the "There is" of presence, even less the "I am here!" of subjectivity, but perhaps the meditative or speculative "Without me!" For, after all, the challenge is indeed to think and experiment cinematically the possible modes of our absence, here less in its absolute form (thinking the world before and after human consciousness)[26] than in the relative form of our withdrawal. To allow anything to make film, Perconte is not outside the shot (extracting oneself from the world is the very movement through which we create nature) but withdraws into the shot, becomes a thing among things. We have seen that long focal lengths and fixed shots are strategies for withdrawing into the shot, being a thing among all those that the film includes and thus allowing all co-actors the possibility of making world and film. Another modality of the human withdrawal in the film comes through the repeated compressions of data which, by thwarting the algorithms programmed to create images reflecting human experience, bring forth the idea of a world-making in which the human audience and their expectations of meaning are peripheral. Indeed, over the course of digital translations of data produced by compressions, it is a non-human logic, directed neither by nor towards the human, that begins to make sense from the data collected during shooting. As Bruno Bachimont reminds us:

> Algorithmic or computational manipulation is only possible if it focuses on discrete elements devoid of meaning. Indeed, … the formal is defined by the expulsion of meaning and sense: calculation is only possible because it does not take into account meaning, but only the syntactic category of the physical symbols manipulated.[27]

By altering the syntax, that is, the rules that allow signs without signifiers to produce effects of meaning, to manage the forms, movements, and time of the film, Perconte lets his compressions produce, from the data collected during the shoot, some modes of being and becoming, interactions, synergies, which do not integrate our expectations of meaning. By not calculating us, this cinema allows us to experience how reality can vibrate when we withdraw from it, in other words, to glimpse a few minutes of pluriverse.

The cinema after nature of Perconte, which challenges our position as absolute subject and considers the agency of all things, poses us an ethical question. It urges us to think of agency not at our scale, but at that of the networks through which we make world with birds, falling leaves, windscreen stains, ideas, waste, dreams, and viruses.

Notes

1. Philippe Descola, *Beyond Nature and Culture* [2005], trans. Janet Lloyd, Chicago, The University of Chicago Press, 2013.
2. Bruno Latour, *Facing Gaia: Eight Lectures on the New Climatic Regime* [2015], trans. Catherine Porter, Cambridge, Polity, 2017.
3. Donna Haraway, *Companion Species Manifesto: Dogs, Humans and Other Partners*, Chicago, The University of Chicago Press, 2003.
4. See notably how Philippe Descola and Bruno Latour link the invention of the concept of "nature" and that of perspective to the objectification of subjectivity. Ph. Descola, *Beyond Nature and Culture*, *op. cit.*, pp. 57–63 and B. Latour, *Facing Gaia*, *op. cit.*, pp. 16–20.
5. Latour uses 'faire monde' in the original text, translated as 'worlding' in the English translation. B. Latour, *Facing Gaia*, *op. cit.*, p.35.
6. Such a perspective can be expressed as follows: "a machine and bodies share a duration that is made of their conjunction. One must then expose oneself, take risks, rub against the other." In Jean-Louis Comolli, *Corps et cadre. Cinéma, éthique, politique*, Lagrasse, Éditions Verdier, 2012, p. 242. Unless otherwise noted, all translations from original French are by the author.
7. "In digital cinema ... there is no such thing as a still image, no punctual moment. There is only a constant process of *becoming* (and *unbecoming*) based on the binary sequencing of zeros and ones that creates a constant relay of appearing and vanishing." In Anthony Bryant & Griselda Pollock, *Digital and Other Virtualities: Renegotiating the Image*, London, I.B. Tauris, 2010, p. 8.
8. Ángel Quintana, *Virtuel ? A l'ère numérique, le cinéma est toujours le plus réaliste des arts*, Paris, Cahiers du Cinema, 2008, p. 46.
9. See the idea of "post-continuity" theorized by Steven Shaviro, who argues that digital cinema no longer treats the representation of space–time from the perspective of the presence of a human

subjectivity, in the work of Steven Shaviro, *Post-Cinematic Affects*, Winchester, 0 Books, 2010. Shane Denson calls "*discorrelated images*" the digital images that no longer correlate with humans and their modes of presence in the world, their capabilities, and limits in Shane Denson, *Discorrelated Images*, Durham/London, Duke University Press, 2020.

10. The concept of flat ontology is developed by philosophers like Graham Harman and his Object-Oriented Ontology (OOO), or Tristan Garcia who uses the term to refer to "theories that do not order worldly entities hierarchically – either in accordance with the substantiality of entities, or based on transcendental principles – but that attribute an equal ontological dignity to each individuated thing". (Tristan Garcia, *Form and Object: A Treatise on Things* [2011], trans. Mark Allan Ohm and Jon Cogburn, Edinburgh, Edinburgh University Presses of France, 2014, p. 11).
11. Nicole Brenez, "Atlantis", in N. Brenez, Ch. Lebrat (eds), *Jeune, dure et pure ! Une histoire du cinéma d'avant-garde et expérimental en France*, Paris / Milan, French Cinematheque / Mazzotta, 2001, p. 11.
12. The term thus typed comes from Lucie Roy, "A digital ontology or aesthetics?", *Sens Public*, 2017, p. 6. http://www.sens-public.org/article1281 [consulted on 17 July 2021].
13. Many of his films are accessible on http://jacquesperconte.com [consulted on 17 July 2021].
14. B. Latour, *Facing Gaia*, *op. cit.*, p. 36.
15. *Ibid.*, note 66.
16. Vincent Sorrel, "'L'alpiniste est un homme qui conduit son corps là où, un jour ses yeux ont regardé': les longues focales de Jacques Perconte", *Le Cahier Louis Lumière*, no.12, June 2019, pp. 85–94.
17. *Ibid.*, p. 89.
18. *Ibid.*, p. 91.
19. The no-matter-what is a central concept of flat ontology that Tristan Garcia develops in *Form and Object* and refers to the potential of existence as a thing of everything, regardless of the rankings and distributions to which one can subject them (natural, cultural, true, false, (im)material etc.). According to Garcia: "Anything can be something, and there is no thing that is more of a thing than another, no thing that is less of a thing than another. Everything

that is a thing is equally so." In T. Garcia, *Form and Object, op. cit.*, p. 61.

20. https://www.jacquesperconte.com/oe?235 [consulted on 17 July 2021].
21. https://www.jacquesperconte.com/oe?28 [consulted on 17 July 2021].
22. https://vimeo.com/171289719 [consulted on 17 July 2021].
23. See for example Peter R. Grant and B. Rosemary Grant, "Unpredictable Evolution in a 30-Year Study of Darwin's Finches", *Science*, no. 296, pp. 707–711.
24. See Baptiste Morizot, "Le devenir du sauvage à l'anthropocène", in Rémi Beau, Catherine Larrère (eds) *Penser l'Anthropocène*, Paris, Presses de Sciences Po, 2018, pp. 249–264.
25. Gilles Deleuze, *Cinema 1: The Movement-Image* [1983], trans. Hugh Tomlinson and Barbara Hammerjam, Minneapolis, University of Minnesota Press, 1986, p. 56.
26. See for example the work of Grégory Chatonsky, http://chatonsky.net [consulted on 17 July 2021].
27. Bruno Bachimont, "Du texte à l'hypotexte: les parcours de la mémoire documentaire", in Charles Lenay (ed.), *Mémoire de la technique et techniques de la mémoire*, Toulouse, Érès, 1999, p. 200.

PART IV

Expanding the Experimental Field

CHAPTER 20

Being with the Birds: The Digital Explorations of Jacques Perconte

Elio Della Noce

We hover over a strange coastal landscape, pierced by electronic fissures; to our left the crest of a cliff, below the assaults of the waves hit the rocks in a surreal manner; with a lateral movement we now approach the wall, on the verge of crashing; but the shot freezes, and everything, mineral, sky, sea, everything becomes painting. These are the latent images on which I ruminate and that led me into the work of French visual artist Jacques Perconte. Even today, their kinetic magic transports me. "Dépaysages [Unlandscapes]" is the title of the series of live performances in generative video from which these images, shot on the Portuguese archipelago of Madeira in 2011, are extracted. And it is indeed through

Interview with Jacques Perconte, by Elio Della Noce
November 15, 2021

E. Della Noce (✉)
University of Toulouse-Jean Jaurès, Toulouse, France
e-mail: elio.dellanoce@live.fr

E. Della Noce and L. Murari (eds.), *Expanded Nature*, Experimental Film and Artists' Moving Image,
https://doi.org/10.1007/978-3-031-70729-2_20

this shift, first semantic, that the digital explorations of the filmmaker touch on the question of aesthetic decentring. To "unlandscape", interrogate the limits of landscape representation, test them, deconstruct them, to "unlandscape" more than to disorient, flee exoticism to find, perhaps even extend, the idea of nature.

Exploring the Mountain

Elio Della Noce: *Does exploring nature with digital tools involve a physical commitment for you?*

Jacques Perconte: My story is very much linked to the emergence of digital tools and their democratization. I was in my twenties and I dreamed of cinema but the images could no longer be made in the same way. Something was disappearing. I really felt that the disappearance of nature, of a sensible world because of synthesis, of mathematics, was an extreme violence, and that the images produced from that moment could no longer be in continuity with our relationship to the world. A world arrives that is produced, that is invented from data, where there is no longer a physical link with the world we have witnessed. Images now function through the link of memory. There is a disappearance, there is an invention, it is much more than the idea of death that we find in photography.

When I started working on the question of landscape my first reflex, which was a reflex born of urgency, was to give the camera to a friend. Returning from an excursion, he filmed the road and the undergrowth, which resulted in the film *Uaoen* (2003). The first lesson I learned from this work was that I absolutely had to shoot my own images and that, in a certain way, these images had no meaning for me. They had to originate from my gesture, my desire. Following this, I started to go into the landscape, into nature. This is all the work that comes around *Uishet* (2007), shot along the Huchet stream in the Landes. But, at that time, I still thought I had to make "good images", that I had to produce things that corresponded to what I could imagine. I had to wait for the right weather window to find the ideal light, and I often spent a large part of my stay not filming. Little by little this idea was constructed that my shoots did not need particular weather conditions and that they in fact did not need anything other than me, it was me who had to change. I had to align myself with the shooting to be able to create the images that only I could create. I realized that I had no idea of what I wanted to do, that there was a question of the gaze of nature and that it was necessarily conditioned by not wanting to do anything. In fact, to really see things, I had to not want to see anything in particular.

There is a first phase of this awareness that continues until 2010, a very exploratory phase. The defining moment of this thinking is the shooting of *Impressions* (2012), shot in Upper Normandy between 2010 and 2011, a film where for the first time I really take a long time to film. I then asked myself the question: "to what extent do I want to let nature do things?" I understood during this shoot that my exploratory mode is an extremely humble mode. When I try to make efforts to go and look for something, I won't get it, while when I make efforts to not look for something, I will see extraordinary things. Vincent Sorrel has been following me in the Alps for several years[1]: he speaks of how surprised he is that sometimes I can simply head towards a tourist viewpoint and set my camera up there. I will film something that, most of the time, looks nothing like what people might film, because I don't really use the viewpoint to film what the viewpoint is made for. It's just that a viewpoint is generally a good place to see.

Elio Della Noce: *I would like to continue on this idea of the viewpoint, of the right place to see the mountain. How did this "ecology of attention" come about in your cinematic exploration of the mountain?*

Jacques Perconte: It happened gradually. At first it was about getting closer, really going to the heart to see, but I realized that, in fact, the closer you are the less you see. From afar, there is so much to see. When I go into the mountains it very often happens that I spend a whole day in the same place. I will walk a bit, sometimes more, with all the equipment, and settle down somewhere. What the mountain offers me in its immensity means that I don't need to move. There is a need, on the contrary, for immense attention on my part to be able to see. Little by little one knows, then one recognizes this or that thing. We saw one thing one day, the next day we see it a little better or differently. This creates something precious in the gaze, for oneself, it's a feeling. Little by little, I see what I film. I have to forget the images that are not mine. I am looking for a particular intensity in the mountain, like at the sea or in the forest. It is born from my relationship to the situation.

There is also an important question of the scale of shapes. The close-up very rarely appears in my work, or it is inexistant. Unless everything is a close-up. I film "wholes", I film sets and when I film the mountain, I film a set of stones, of rocks, I never film the mountain as a unit either but as a set of elements that are not separated from each other. For me, the mountain's most important notion is that of the off-screen. The mountain necessarily exceeds the frame. As it is impossible to film in its entirety, what I show of it describes the lack, the part of a whole that is missing.

Elio Della Noce: *An important part of your works gravitates around the mountain: I am thinking in particular of* Les Alpes *series or* Avant l'effondrement du Mont-Blanc *(2020). Where does this fascination for mineral come from?*

Jacques Perconte: On this point, I would say very clearly that I make no distinction between the mountain and the sea. I see mountains as waves, except they are not waves of water, they are waves of stone, it is only their speed that changes. The power of the wave is the force that raises the sea and at the same time the water that collapses around it. There is the return of the water, this movement that brings the water back. We are caught in this movement. I greatly enjoy this about the sea and the mountain. In that instance, there is the tectonic pressure which makes the mountain grow, the glacier melts, the mountain grows and collapses. This is the natural movement. My film *Avant l'effondrement du Mont-Blanc* (2020) particularly focuses in one sequence on climbing the slope that collapses, on climbing the tangle of stones that detach and are going to fall (Fig. 20.1).

I do this very often. For me there is a vertical movement of the camera that gives the impression that the mountain is descending while my gesture is rising. The mountain is only that, it is only this double movement at such a slow speed that it is hard to see.

The relationship I end up having with the mountain does not come from a dialogue but from my own feelings. It's an inner journey towards stillness. When I connect with the mountain, the mountain does not speak to me, I welcome what it radiates. It's the contact with the wind, the contact with the stone, with the light, the smell. It's all the senses. But it's also inner calm, that is, the absence of commentary, and therefore a huge conceptual detachment, the greatest possible. My reference on this subject is that of Eckhart Tolle, who is a spiritual master. One of my favourite books by Tolle is *Stillness Speaks* (2003). For me, the question here is that of stillness, both the inner one of my feelings, of my calm, of my well-being, of my humble comfort at the extreme opposite of the conqueror, who is comfortable because he thinks he dominates, that he wins over himself and the elements; and that of not desiring much, of being in what is really happening and therefore of truly being able to

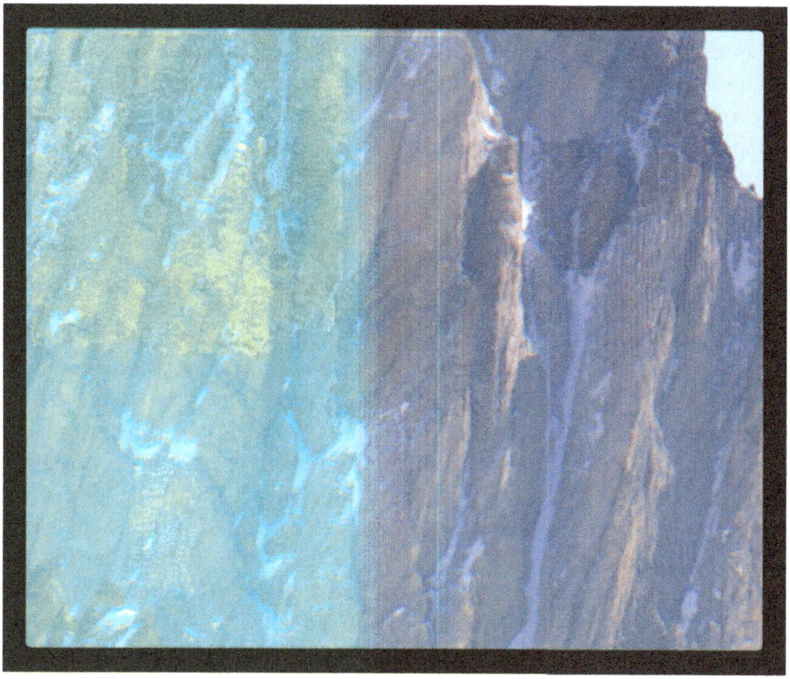

Fig. 20.1 Jacques Perconte, *Avant l'effondrement du Mont-Blanc* (2020)

see. There, I feel the fact that the mountain exists in a different space–time and that its movement is perceptible, that it is perceptible in the writing of its forms and that my gesture can naturally find its way. This is something that does not necessarily happen in thought but in the feeling of a landslide, the following of a force line in the rock.

To take the example of *Avant l'effondrement du Mont-Blanc*, we have these two laterally opposed elements in the film. On one side we have the moment of quiet which is when I let the plane frame the image entirely; when I am with the camera in the cockpit I do not try to film; trying to resist the movements of the plane caught in the hot and cold air currents, all the turbulence, would be doomed to failure; I let the plane decide the frame of what will be in the image, my attitude is not to look in the viewfinder, I simply look at the start, there is Mont-Blanc in the image, then the plane does the rest, I hold the camera. The question of the collapse of the mountain carries an *economic dis-quiet*. If the mountain collapses, it brings down ski resorts and crushes villages. This prevents an economy. The permafrost is disappearing, there are many mountains collapsing but we are not particularly worried about it. We are only very worried about those we actually see because they are the most frequented for hiking, skiing, and mountaineering.[2]

Seeking the Living Being of Images

Elio Della Noce: *In your films, I get the feeling of accessing a "wildness" of natural environments, of the mountain, the sea, the sky, or the forest, as if your images have renewed the question of the wild.*

Jacques Perconte: For me, the notion of "wild" is what separates landscape from nature. Wildness is easily present everywhere but very often so minutely that it is difficult to grasp. I think one of the challenges of my work is to go and find this wildness in any situation, to produce an image of nature that counters an image of landscape. Very often, by letting the wind make the camera vibrate, by letting an apparatus frame in my place – a plane, a boat, a train – the vibrated image will allow things to emerge. The tool is put in danger (sometimes it breaks), the technology is put in danger, normally there to stabilize, optimize, neutralize, dominate. In my images, I recount these exceedances, this quest for a liberation of contingencies.

This loss of control is perhaps what brings back wildness. It's when technology no longer allows us to protect ourselves from nature, when we are in danger. I work a lot at this frontier, letting things happen, by exploring the control capabilities of my tools, so that the freedom of forms can emerge from the images, so that we realize, truly, what is living and what is technological. I'm not trying to oppose these dimensions, but to bring them together consciously. Video compression, to optimize the portability of streams, endangers the readability of images. To maintain visibility, it will neutralize the discordances of this mathematical universe and make it disappear to give the illusion of something photographic. Among the matrix-like computer structures, forms that remind us of the living appear, that are truly alive, not the image of a living thing, but living things in their organicity, in their way of moving, in the agency of elements in relation to each other, in the contingencies of encounters between forms and rhythms. We recognize in the images what makes us think that something is alive.

Elio Della Noce: *Indeed, in your images, I witness a tension between the forces of the living and what I feel is a mathematical fixity. Does your use of digital processing tools lead to the idea of an "image biomorphism"?*

Jacques Perconte: My research consists in finding how traces of the living can remain within the images. To recount this balance, this tension, to the point where it is no longer a separation but a continuity. My projects continue the filming experience through another experience which is that of the image, in which there is a persistence of something living, which no longer shows the subject in a defined way. My desire is to produce images in which we are connected with something that is of the living, and that we connect as a spectator to this living in something that is not living, or is not in the same way.

In the middle of *Ettrick* (2015), one of my films about Scotland and in which the subject is forestry, there is a whole set of forest scenes. These are fixed shots on woods with almost no wind. When the images move from one to the other, we simply see a few leaves moving, a trunk slightly appearing when a bit of light has flickered, but we are always in the presence of the previous image. The novelty of the image is just the manifestation. The visual manifestation here is the fact that the image gradually completes itself, that we recognize the forms, attribute the manifestation to the form and associate the movement with the object. The idea is to move towards this question of manifestation, of something that is there "outside the image" but "through the image".

Painting the Sky

Elio Della Noce: *I would like to discuss now your triptych* Or: Or/Aor, Budapest *(2018)*, Or/or, Hawick *(2018)*, Or/Aour, Wien *(2019). These three films share the close-up image of a canvas (a golden background) that opens to a sky where birds are twirling. Here you give the impression of expanding a form of living that painting would have "condensed".*

Jacques Perconte: When I film birds, the desire is to be fully with the birds, without telling myself: "I am a bird, I fly with the bird", not in a naive way. I find that to see birds playing in the sky is one of the most beautiful spectacles, firstly because there is of course the dream of being able to free oneself from gravity, to fly, to be free as birds are free, and then because there is something magical about birds. For me, they are the true intercessors between Earth and Heaven.

When I film birds, I try to fully tune in to the movement they make in the sky. More specifically *Or/Aour, Wien*, I fly, well, I play with the birds. The camera films the swallows in the sky. I dream of going with the swallow without following it, that is to say, as if I were running with someone, feeling their run, being in a mode of coexistence, a form of presence to what is happening, which makes me in tune with the moment. There is a moment of meeting between me and the flight of the swallow. I want to banish the reflexes of the type: "she will escape me", "I need to speed up the camera", or "she will turn", "I need to anticipate her movement". The story of my films is very much a search for this erasure of dictates, to be able to produce a flow in which I am carried away, where I lose the notion of time, where I whirl, where I fly.

This "Gold" series is both a relationship to painting and to alchemy. Very literally, this series is "dipping birds in gold". And the gold that I wanted to work with at that time, I wanted it to be a "gold imbued with love". For me, that's *The Kiss* (1908–1909) by Klimt. So I went to take photographs of the gold leaf on Klimt's canvases. The film is a slide between the photograph of a detail of Klimt and the shot in which there are the birds. The gold canvas is actualized by the movement of the camera and the birds in the sky. The colours are carried away and these are the colours of this love-gold that are taken by the movement and the story of the birds (Fig. 20.2).

Elio Della Noce: *In* Or/or, Hawick, *there is a moment of pure pictorial materiality: hundreds of horizontal lines end up overlapping on the gold of the sky and forming a kinetic weave of bright colours, drifting from right to left.*

Jacques Perconte: Very early on, the challenge for me was to discover a pictorial technique specific to the digital image and in particular to moving images. This technique would come from a moment when I started to work images with computers and where it was not possible to do otherwise than to work with compressed images. Quite quickly, I realized that compression was not something I should consider as a flaw but as a potential. The way I used these tools was hugely influenced by painting in terms of technique. It's my culture. The quality of the image meets the quality of the tools we use, which produces a sensation of presence, something material. There is an inclusion. My images show something of their screens that other images do not show. Each one is

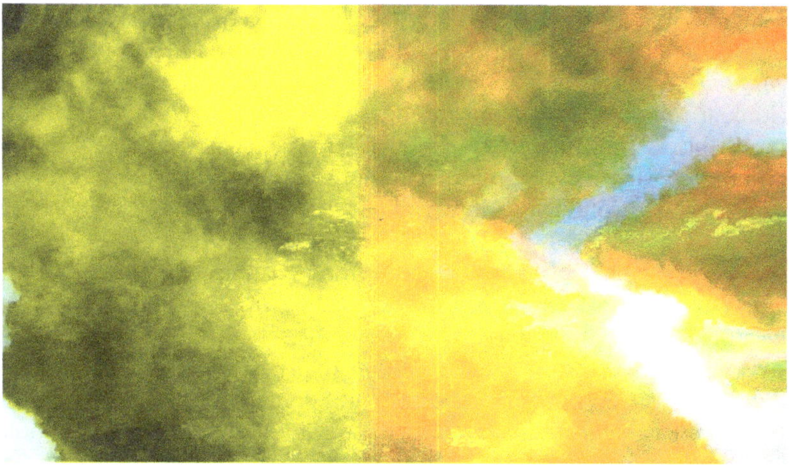

Fig. 20.2 Jacques Perconte, *Or/Or, Hawick* (2018) © All rights reserved by the artist/Courtesy of Light Cone

made for a particular apparatus. They show the infrastructure, the colorimetric power, since my colours use the maximum capacity of the screens. They show scales of contrasts that we never see, because the contrast scale of my images is the matrix scale of the screen.

Expanding the Horizon of One's Practices

Elio Della Noce: Salammbô *(2021) is a film performance created for the bicentenary of the birth of Gustave Flaubert. The historical novel seems to lead you to images of the Anthropocene: factories, chimneys, smoke, wild animals. How does the collaboration with live performance influence your relationship with images here?*

Jacques Perconte: I'm careful with words like "Anthropocene". I keep them out of my discourse, and generally, it's not me who uses them. In the introductory sequence of *Salammbô* there is a mix of images from the Port of Rotterdam and images from Normandy but I am not denouncing the violence of this situation of the Netherlands' industrial domination over the environment. When you live in Rotterdam, you feel the enormity of the Port, with its 42 kilometres of port infrastructure. Ten minutes from my home, in the city centre, there are docked offshore oil platforms. The absurdity of this enormity vibrates the industrial sublime. In *Salammbô*, I feel more like I'm in a relationship of documentation, where I look at things, without showing or denouncing them. A chimney that smokes black, I'm not going to show it because it smokes black. I'm not trying to construct something that argues something. It's very difficult to extract oneself from a situation where you build an ecological argument (Fig. 20.3).

The piece *Salammbô* has two distribution channels: it is made for the stage but has also been shown in the cinema. In fact, I would say that the difference between these two distribution modes lies in the way I envisage the space of representation. For the stage, the image anticipates in its frame the place of the musicians and lights them. The image must be presented in a certain way, in certain dimensions. There is a whole

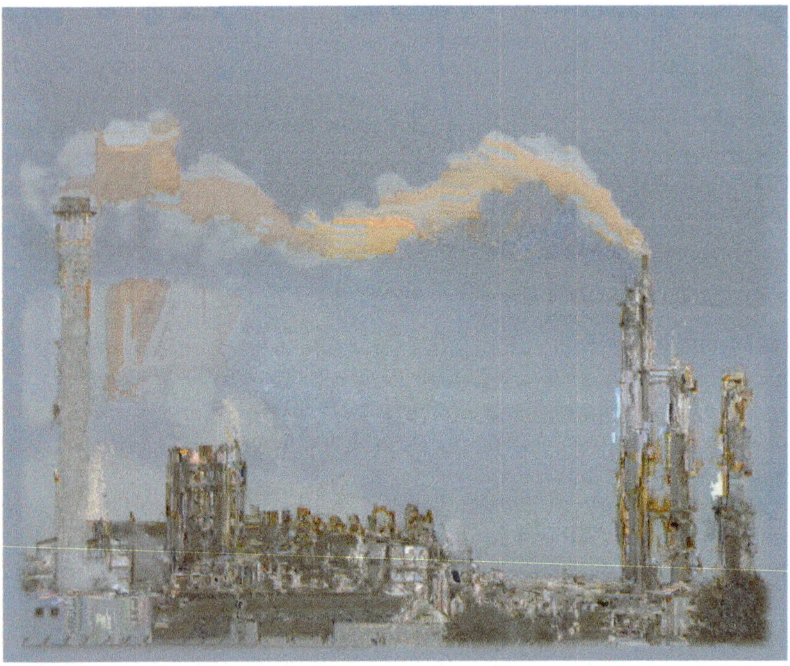

Fig. 20.3 Jacques Perconte, *Salammbô* (2021) © All rights reserved by the artist/Courtesy of Light Cone

stage context that is much more precise for the presentation of the performance, whereas in cinema we are truly in a relationship with the content of the image and the music in an immersion that is that of the movie theatre. For me, these are different ways of showing the piece.

At the centre of my practice, there is a kind of historicity of images, which are first exploited as installations for exhibitions, are then performed, and finally they stabilize or become the core of a film work. Then, there is also a purely experimental dimension, that of seeking the place where my images function best, not to find it but to experience it, and to have different relationships with audiences, to share other forms of energy. *Albâtre* (2018) is, just like *Salammbô*, a film designed to be shown with a group of musicians, while *Faust* (2017) is an opera only, without a version for the cinema. When I was working on the project *Seconde nature* (2020) with choreographer Fabrice Lambert, it was about posing this question of human presence on a stage in front of a screen again: how to operate this extension of cinema, how the image that one might believe is behind, in the background of the stage, can integrate bodies, work with them on the same level as the sound dimension, to make everything unified without artifice or devices.

Elio Della Noce: *Does the idea of an "environmental" aesthetic, which relates to your practice of a cinema of nature, not find a form of continuation, a shift, in your practices of expanded cinema?*

Jacques Perconte: For me, this is true for everything, for all films, even those that are not mine. It's the idea of an environment that is created by the space–time of an experience. There are more specific questions that I'm asking myself for an upcoming exhibition, those of bringing natural elements into the exhibition space, stones, things like that, because I am very attached to mineralogy. I work a lot with natural materials. In the piece *Europaaour* (2022), I work with gold leaves and stones, for example. In these cases, there is a continuity between the environment and the projection space, since the presence of a physical material in the space contributes to the perception of the image. Indeed, one could imagine talking about an environment or ecosystem but these remain difficult concepts for me. For me, an ecosystem is not something that is made by man, an ecosystem is something that is autonomous, with actors who are autonomous but who work together.

Elio Della Noce: *Jacques Perconte, would you say, like filmmaker Pip Chodorov (starting from his experience of photochemical practices and* *artist-run film laboratories), that you are not in an economy but an ecology of image production?*[3]

Jacques Perconte: Yes, I could say that, but I think that in my texts you don't find "ecology" but "economy" because I don't like to oppose these two terms. For me, economy is not a bad thing in itself. It's perhaps my relationship with tools that makes me say this. Most of the time no tool is bad in itself, it's the destination and the use that poses a problem. We shouldn't forget that with "economy" we say "economical", that someone who is economical is someone who does not waste anything. [NB: "*S'économiser*", "to economize oneself" in French is an expression which means also to save oneself, save one's energy, for later.]

I do not oppose industry and nature. We exhaust resources but we don't destroy nature. We exhaust a visible dimension of resources. Perhaps there are planets full of these minerals that have become rare on Earth and represent multiple infinities of the quantities we would need in the total life of humanity, if it ever ends. Can we or should we exploit them? I imagine that humans will never be reasonable and that the idea that I love so much from Ivan Illich, that of the "necessary maximum",[4] will never really be embodied. The notion of nature represents for me an infinity. It's really a set that evolves by compensating creation/destruction just like planets, which are born and disappear. This is how nature works, and we humans, we still consider nature as something to save, to modify, to evolve. In fact, it's only our living environment that we can save. So this situation of the human, of how he thinks and builds an idea of ecology, is strange for me, since an ecology cannot be compromised, made of compromises, otherwise it's just politics. We reach a plateau that even a politics of degrowth will not be able to decrease, which is the minimum standard of living of our industries.

Notes

1. Vincent Sorrel, a filmmaker and researcher in visual studies, has notably written on the Alpine explorations of Jacques Perconte, including "'L'alpiniste est un homme qui conduit son corps là où, un jour ses yeux ont regardé': les longues focales de Jacques Perconte", *Le Cahier Louis Lumière*, "Arts filmiques et expérimentations optiques contemporaines", no. 12, June 2019. Available online: https://hal.univ-grenoble-alpes.fr/hal-01991642.

2. See the presentation note of the film *Avant l'effondrement du Mont-Blanc* on the filmmaker's website:

   "Are we the last ones able to see the peaks of Mont-Blanc? The heat of the summers, the too mild winters are largely responsible for the rock collapses, which have been multiplying for about twenty years. The mountains are collapsing. If it's a sign of climate disruption, it's also a sign of our attachment to the landscape that we would like to be able to classify as heritage. The Mont-Blanc massif is not ours, the mountain is a state, it's a moment, it was not, and it will change anyway. The problem here would be that of speed. Because the balance of these peaks that defy the void, the longevity of these glaciers is only our point of view. On the scale of the movement of the planet, it's a vibration. The mountains fall, we can do nothing about it. And even if we have the means to rise to their height to admire them, to surpass these inaccessible peaks where many explorers have lost their lives wanting to access the privilege of conquering them, the mountains will continue to fall as they continue to rise. If Mont Blanc collapses, it also rises." https://www.jacquesperconte.com/oe?244.
3. Pip Chodorov, "The Artist-Run Film Labs", *Millennium Film Journal*, Vol. 60, Autumn 2014, pp. 28–37.
4. Ivan Illich is a philosopher and thinker of political ecology. Regarding the "necessary maximum", refer to his work *Energy and Equity* [1973], London, Marion Boyars Publishing, 1974.

CHAPTER 21

Diving into the Deep End, to Renew and Extend Our Perceptions

Vincent Deville

There exists in cinema a figure that I propose to name the "leap into nature", suggested to me by Jean Epstein's *Mauprat* (1926). In this feature film, the aristocratic Mauprat family has split into two radically opposed camps: one lineage attached to the values of nobility, while the other is composed of ruffians. While Bernard de Mauprat, as a child, was kidnapped and raised like a "wildling" by the outlaw branch, he is reclaimed as a young man by his cousin Edmée's family, who try to instil in him the values of good society. But the young man nostalgically recalls his previous life. Overwhelmed by boredom, in a plush interior where nature has been entirely reified (in the tapestries, clothes, and so on), Bernard suddenly and furiously tears off the fine clothes he has been dressed in to delightfully rediscover his old rags. Leaving the house with determination, he literally leaps and bounds into the wild. Epstein marks this transition from civilization to undomesticated nature with a low-angle

V. Deville (✉)
University Paul Valéry Montpellier 3, Montpellier, France
e-mail: vincent.deville@univ-montp3.fr

E. Della Noce and L. Murari (eds.), *Expanded Nature*, Experimental Film and Artists' Moving Image,
https://doi.org/10.1007/978-3-031-70729-2_21

shot, Mauprat jumping towards the camera. The cut to a tracking shot that accompanies his run and the shots of a delighted Mauprat's face, who seems to move without touching the ground, a flower between his teeth, leave no doubt about the ecstasy of freedom regained in this act of rebellion against a corseted society. Epstein invents here the figure of the "leap into nature", which consists precisely in bringing together the human and the living that surrounds and constitutes him, as soon as a film character has been separated from his natural environment by an artificialized society, and he regains this original environment in a performative leap. One could trace within narrative cinema the history of this filmic figure based on the arrangement of motifs and forms (the leap itself, the torn clothes, the run, the transition from architectural forms to natural forms, the cut from a static shot to a tracking shot, the speed). It is also possible to transpose this history into the gestures, techniques and processes of experimental cinema.

This is what I propose to do here by considering the leap into nature as a dive: when the human implements an aesthetic perception that immerses him in nature; when a body is itself immersed in a natural environment, giving rise to forms of entanglement; and when the very processes of image production resemble a dive into the deep end, especially to reveal analogue images produced by ecoprocessing. These different manifestations of human relationships with nature will be considered within a corpus of films, mostly recent ones, thus highlighting the prevalence of this concern among contemporary filmmakers.[1]

Perceptions

In his book *The Veil of Isis. An Essay on the History of the Idea of Nature*, French philosopher Pierre Hadot distinguishes three different ways of relating to nature in our human experience: the *daily* and usual perception; the *scientific perception*, which modifies our knowledge without necessarily modifying experience; and the *aesthetic perception*, a perception "for pleasure", when things are seen for themselves and not for oneself. It is this gaze that interests us here, which Henri Bergson summarized as "contemplating the universe with the eyes of an artist",[2] and which already induces a decentring, inviting us to reinvent our perception beyond the usual and rational, in order to produce a new understanding of the world.[3]

Jonas Mekas gave us a magnificent example of such perception at work in *Walden* (1969) when he visits his friend Stan Brakhage, who lives with his family in a house in the middle of the forest. From very early on and throughout his prolific career, Brakhage pushed human and cinematographic perceptions to the limit. Mekas observes him in an apparently trivial moment: Brakhage listens to music, lying on his bed with his dog, observing the forest through a large window placed right in front of the bed. I have often thought that there is no better image of the artist at work! One can imagine that Brakhage himself designed this space as an artist's studio: a bed placed in front of a bay window opening onto the forest. Mekas translates his act of vision, that is, the transition from inside to outside, by first framing the window, the dog and the filmmaker, then he "leaps" into nature by now filming in long focal length the tree trunks and more particularly a bird. We have moved from the body of the observer to the observed images. What strikes one again here is the connivance between idleness and the work of the artist, as Hermann Hesse very well described it:

> But artists have always needed temporary idleness, partly in order to let newly acquired things clarify themselves and unconsciously working things mature, partly in order to approach the natural again and again in careless devotion, to become a child again, to feel again as a friend and brother of the earth, the plant, the rock and the cloud.[4]

In his film *White Out* (2021), Chris Welsby extends Stan Brakhage's observation gesture. One morning, Welsby films through a window fir trees under the falling snow. Three modifications distinguish this perception from a natural perception: the film speed is slowed down; the brightness is varied, alternately lightening the image to a white that invades the screen or darkening it, highlighting the fir trees under the snow; a white noise accompanies in sound waves the waves of images, crescendo and decrescendo. When the filmmaker writes: "I was thinking about what one feels when watching the snow fall", it reflects the subjectivity of this perception, but he implies a shift: as if the aesthetic perception, in making the visual and sound waves coincide with the snow squalls, operated a shift from human perception towards the meteorological phenomenon perceiving itself.

This specific quality of aesthetic perception, which is based on the choice in the field of vision of what could have been left aside, has three key characteristics: a change of register from "cultural" to "natural", working towards a profound modification of our anthropocentric vision; a change of shot scale, generally in favour of looking closely; and a study of details, attention to things that are usually beyond ordinary vision, a gesture which, by dint of its proximity, could lead to a leap into abstraction. In this respect, Ralph Steiner's *Look Park* (1974) constitutes a real lesson. The film begins as a visit to the aptly named Look Park, and the introductory card explains: "This film of a park waterfall was made without any camera or lens tricks. The reflections of trees and sun may seem fantastic, but they are what anyone could see who looked closely." However, it is the filmmaker's aesthetic gaze, and therefore his sensitivity to the world, that gives us access to these extraordinary visions. Indeed, once the wide shots contextualizing the park have passed, accompanied by bucolic Offenbach music, we witness a leap into nature in the form of a dive into a parallel world, which takes on the appearance of "a leap from quality to quality"[5] based on a series of transitions: from wide shots to close-ups on the water; from figuration to abstraction; from colour to black and white; from classical to electronic music. We no longer see a world of appearances but its reflections which seem to contain its deepest truth, making us move from a ***natured nature*** to a ***naturating nature***, from the produced form to its productive principle[6] (Fig. 21.1).

An abrupt transition takes us back to the world of ordinary perception. This first perceptive dive is a foretaste of an even deeper need for decentring.

Entanglements

French researcher in plant ecology Jacques Tassin observes that human beings have become estranged from their original living environment, and that this estrangement is deleterious – in the primary sense of endangering life. This is reflected in exponential urbanization and the eradication of natural environments:

> How have we become strangers on our Earth? How were we expelled from the biosphere and how could we let the dialogue with what is not human be broken? But above all, how can we now restore this original conviviality that we lack so much? ... We could have, each of us, very simply opened our senses, reinserted ourselves into this vast and wonderful sensitive matrix

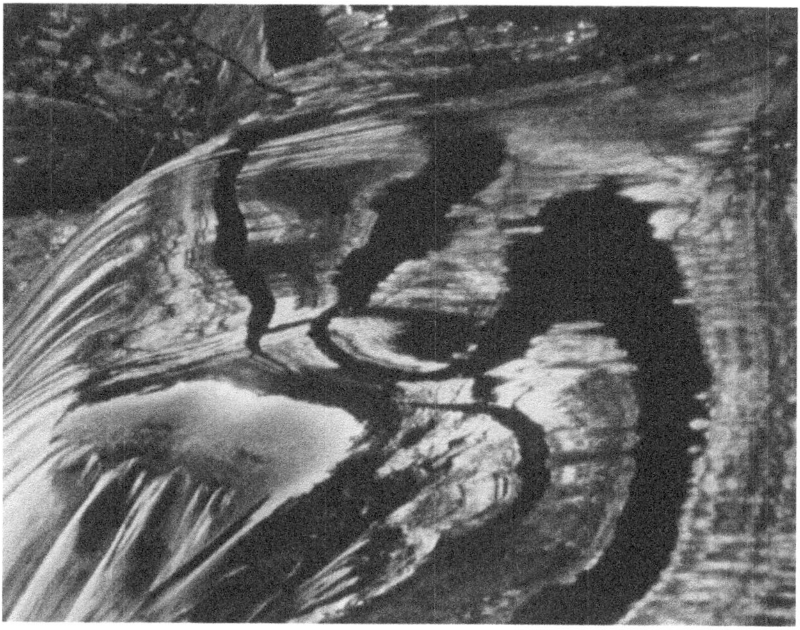

Fig. 21.1 Ralph Steiner, *Look Park* (1973–74) © Courtesy of Bruce Posner

> that immerses us in the continuity of the living. No being can constitutively withdraw from it. Yet, modernity has torn us away from it.[7]

The Canyon (2021) by Zachary Epcar notes this state of affairs while pointing out the need for a decentring of the gaze. From the very first shot, *The Canyon* describes a contemporary residence immediately associated with "high standing". But this perception is observed through foliage, as if we were seeing the scene through the eyes of a neighbouring plant, observing the strangeness and incongruity of human *habitus*. Producing a strong effect of ***defamiliarization***, in the sense of Viktor Shklovski as explained and analyzed by Carlo Ginzburg,[8] this shot immediately introduces a critical distance. The reversed perception of the human by the plant, as a perceptual decentring, is an extremely powerful performative act, which in one and the same gesture perceives, notes, criticizes and proposes a solution. Seeing through the prism of natural elements puts our ways of seeing and therefore thinking into perspective. And this simply by replacing the human in nature, by *reterrestrializing* him, as proposed by Jacques Tassin, echoing Augustin Berque's concept of *deterrestrialization*[9]:

> We need to *reterrestrialize* ourselves. And in order to do so, we must rid ourselves of this idea of a nature separate from us, of this virtual world of which at least part of ourselves would not be a part, according to the rules of a great divide. ... It is a matter, through the sensorial act to which a fully accomplished culture leads us, of becoming fully human again, and therefore just as fully reintegrated into nature.[10]

The Canyon multiplies the signs of a modernity that has continually distanced us from the natural world: balconies all equipped with the same garden furniture; electric toothbrush; plastic cup and straw; mobile phone; wireless earphones; iPad broadcasting images of a waterfall during a massage session.[11] As if to counteract this litany of commercial modernity, the filmmaker opposes another series of shots of the residence taken from a plant perspective. It is a way, as Jacques Tassin writes, of reinserting us, by immersing us into the continuity of the living. Plants are a force of protest against the vanity of human entertainment and maintenance activities (swimming pool, *running*, tennis…), as in the scene of the tennis court where an exasperated player throws his racket in a rage and leaves the court while we are left with the wild grasses climbing up the fence.

Violence here is expressed primarily through the orthogonality of the architecture, which is fundamentally opposed to the free and curved forms of the living. The filmmaker insists on this striation with functionalist rigour when he replaces architectural views with their prominent lines, animated in white dotted lines on a black background. Literally, it seems to refer to the opposition between the *smooth* and the *striated*, as analyzed by Gilles Deleuze and Félix Guattari in *A Thousand Plateaus*, from which we can retain some ideas that shed light on the reading of Zachary Epcar's film: an opposition between open, undelimited space, and the *polis*, the city, the town; but also a striated city that would rework smooth space within itself, wherever possible, which occurs here in the reappearance of vegetation wherever it had been constrained. And what the two philosophers write about industrial wastelands ("The smooth spaces arising from the city are not only those of worldwide organization, but also of a counterattack combining the smooth and the holey and turning back against the town."[12]) can be extended to non-human living beings who, by rewilding architecture, regain ground in space and visibility in the image. Because, as Bruno Latour writes, "animate beings are not limited by frontiers and are constantly overlapping, embedding themselves within one another".[13]

The forms of interspecies entanglement constitute a vast field of experimental and avant-garde cinema. Rose Lowder gave some of the finest examples in her *Bouquets* (1994–), which manages to mix all the terrestrial constituents through an optical fusion of frames. *Sexe-thym* (1993, 0′45) by Nathalie Harran is a much simpler and more modest technical work, distinguished by its spare style, the simplicity of its design and its extreme brevity, without diminishing its poetic impact. The film consists of a shot of a few dozen seconds on a female sex organ, whose frontal framing initially evokes *The Origin of the World* (1866) by Gustave Courbet. The camera performs a zoom in and sprigs of thyme appear blurred in the foreground, intertwining with pubic hair, as if the plant came to ally with the feminine and protect it from the overexploitation of sexual images. In their film *Trahere* (2018), Caryn Cline and Linda Fenstermaker use superimposition to blend a female farmer working with her tractor among the environment of her farm, particularly filled with flowers of many colours. Presenting the film, Caryn Cline speaks precisely of a "timeless relationship with the natural world as a young farmer *entangles* herself among the flowers".[14] In *Purkyně's Dusk* (2021), Helena Gouveia Monteiro explores the variations of colour perception in low light conditions, in reference to the work of physiologist Jan Evangelista Purkyně, who gave his name to the Purkinje effect.[15] The filmmaker poses a man and a woman in a park, alternating shots of them and of plants and flowers, working on the immersion and fusion of humans in the plant world and in the silver film material. The film constantly proceeds by jumps: from plants to humans; from a transparent mimesis to an affirmed plasticity. Processing spots on the film induce the idea of immersion: just as the film is immersed in a chemical bath, the couple is immersed in a bath of nature, greenery and colours, until they blend into it and disappear. On the path to egalitarianism, the end credits indicate after the names of the two human actors the names of the plant actors that appeared on screen: *Papaver, Geranium, Skimmia, Malus, Euonymus, Escallonia, Nymphae, Cordyline.*

Re-terrestrializing the human invites us to re-imagine the whole of the living, in the sense of putting it back into the image, for itself.

Dives

But rather than trying to represent the living for itself,[16] some artists seek to approach natural perceptions, where nature produces its own modalities of perception and figuration. How to approach a perception freed from our anthropocentric habits? By adopting rhythms, forms and textures, or even chemical processes.

In *Washi Wishi Washi Wav* (2020), Miles McKane seeks to match the rhythm of the waves by working on four parameters: jump-cuts edited in-camera; slow and fast motion; an expanded projection that alternates between double and single screen; and a collaboration with a musician who proposes a rhythm-based music. This study of waves first refers to a seminal film for avant-garde cinema: *The Wave* by Étienne-Jules Marey, filmed in Naples in 1891, whose fertile links with contemporary experimental cinema have been well demonstrated by Nicole Brenez.[17] If McKane and Marey share the motif of the wave,[18] they also share a commitment to nature as a source of inspiration, as expressed by the physiologist:

> The study of nature has always been the source from which artists have drawn their inspiration. The higher the art rises, the more respectful it is of the laws of anatomy; thus, we see it, even in its most personal creations, adhering to the faithful copy of the living model.[19]

In *Washi Wishi Washi Wav* the procedure itself produces some "qualitative leaps" which generate rhythmic incidences: in the interval between the two screens, which produce consonances and dissonances between them; each time we move from double to single screen; when we switch in mono-screen from the motif of the sea to a rural countryside scene with a misty atmosphere, and whose contrast is further enhanced by the appearance of a horse that seems straight out of a film by Federico Fellini or Youri Norstein; and when, finally, a last long passage in mono-screen focuses exclusively on the waves getting closer, like a real dive into the waves, which is reminiscent of a return to the *primordial soup* of our origins (Fig. 21.2).

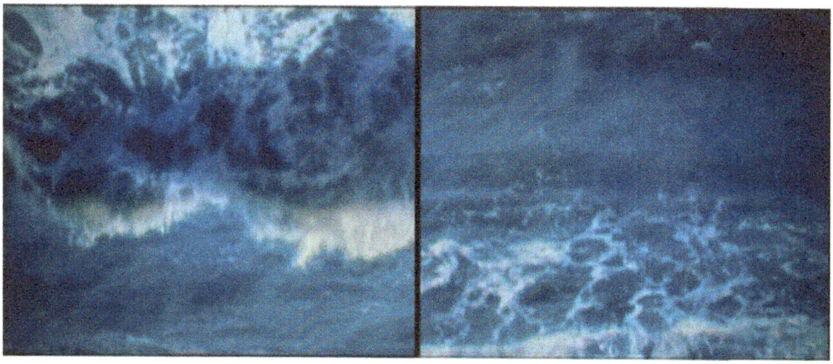

Fig. 21.2 Miles McKane, *Washi Wishi Washi Wav* (2020) © All rights reserved by the artist/Courtesy of Light Cone

For his generative piece titled *The Tempestaire* (2020), which he places in the lineage of Jean Epstein, Jacques Perconte seeks to embrace the textures and forms of waves by working on the deep affinity that is established between a natural phenomenon and pixels. It is the very process used that re-enacts the logic of the formation of waves as a perpetual movement, from the same constantly reconfigured material, as the filmmaker explains:

> [The shapes and colours] are born from what was already present and displayed. The new shot is an actualization of the old one in terms of textures and colours. ... Thus the film is a flow in constant modulation. ... Constantly fed with images by the programme, the film is edited *ad infinitum*.[20]

Another correspondence between image and wave is played out on the edges of the frame, similarly agitated and fluctuating: just as the waves follow the ebb and flow of the sea and reconfigure its edges and limits, the inside of the wave can become the tip and the tip can return to feed the trough of the wave.

Other filmmakers, working in groups where knowledge and know-how are shared, have striven, often at the cost of many attempts, to push the logics of hand-processing towards a non-polluting approach, by replacing harmful synthetic chemicals with natural ingredients. The members of the MTK workshop in Grenoble have pursued this logic by seeking to develop films from plant extracts belonging to the very place where the images were shot. The protagonists of the collective film *L'Amitié ce n'est pas toujours du ski de fond* (*Friendship Is Not Always Cross-country Skiing*, 2021), in order to conduct tests with several plants (nettle, thyme, sage, yarrow) and to measure the potential of their active agents to create processing baths, first conducted a collective shoot in the form of a big bath: during a sunny afternoon, a group of young people, some naked, bathe in a mountain lake, have fun and seem happy with this moment of shared friendship. The results of the processing tests come to be inscribed in the flesh of the images, deeply affecting them. Joyce Lainé (filmmaker) and Jacopo Rasmi (researcher), who both participated in this shoot, explained during an academic event[21] some of the logics at work in this research: mixing people of different origins and backgrounds to exchange their skills but also their incompetencies; developing forms of mutual aid and collective modes of creation; developing new narrative forms that include the procedures of creation in the work; the lack of control of the result to come; bringing together ecological, health, economic and political concerns regarding non-polluting products for users and for the planet, easily accessible and available for everyone, independently of industry; and organizing workshops and screenings in a multitude of places according to a logic of intergenerational transmission (MTK was founded in 1992) and of geographical influence.. Here we could propose extending the notion of mutual aid in the way that the plant is directly associated with human activities. Each plant can thus offer what it is itself capable of, what it is "gifted" for (a power of contrast, colour, texture, grain…). The nettle, a ruderal plant that grows particularly in places that have gone through an extinguished human activity, notably on the ruins of industry, as the filmmaker and the researcher remind us, can be called the agent of a palingenesis, a new birth, proposing new escapes for creation in the interstices of collective and open artisanal workshops.

The *ecoprocessing* movement is, however, not totally new. Philip Hoffman was one of the pioneers in Canada with the Film Farm and his theory of *vulture aesthetics*, which insists on the relationships between community and (film) processing, requiring an openness to unexpected temporalities and their environmental implications.[22] Why this animal metaphor? According to Hoffman, the filmmakers who come to the Film Farm and vultures have certin features in common: a piercing gaze; they seek decaying food, which their stomach acid or their natural processing baths will allow them to digest or reveal; they are the ultimate Freegans (a movement that opposes any paid consumption, in favour of the free use of everything that is thrown away); this recycling allows filmmakers to bring back technologies that have been abandoned for reasons of standardization and profitability; the idea of a shared feast (waste, knowledge, values); and all for the benefit of a "symbiotic relationship with the world, a kind of empathy with natural forces. Allowing these omniscient forces to be in partnership with the making of film work, ultimately puts one in beat with the world".[23]

Through the different ways of soliciting natural elements in the creative process, they move from the status of backdrop to that of agents endowed with the power to act. It is the same dynamics that lie behind the epistemological questioning of environmental history, as Grégory Quenet presents it:

> Historians had to stop being content with an environment taken as a backdrop, a simple framework on which the action of men unfolds. More radically, it was rather a question of considering natural elements as full-fledged actors, endowed with their capacity for action and their own logics, not reducible to the use that men make of them.[24]

This agency of natural elements allows us to glimpse an "extended form of consciousness (*awareness*)", as filmmaker Karel Doing writes about the plants with which he makes films[25]: a statement that aligns with the definition of expanded cinema given by Gene Youngblood:

> When we say expanded cinema we actually mean expanded consciousness. ... Expanded cinema isn't a movie at all: like life it's a process of becoming, man's ongoing historical drive to manifest his consciousness outside of his mind, in front of his eyes.[26]

To give back to all living beings, human and non-human, the keys to their own agency, is to make expanded cinema, is to participate in the construction of an extended consciousness.

Just like Pierre Hadot regarding aesthetic perception, French philosopher Marcel Conche also drew a distinction between what ordinary and scientific perceptions could embrace on the one hand, and what of nature escaped them:

> But is nature reducible to the universe of physicists, to the universe of the Big Bang? Is it reducible to the world? Or should we rather think that the structured Whole that is offered to us does not encompass everything there is, and that there is infinitely more to know beyond? ... The set of things, which we here call "nature", presents itself as never being able to finish giving itself. Thus, meditation on the "Whole of what there is" becomes a meditation on nature, as an untotalizable totality, inexhaustible, undoubtedly infinite.[27]

This idea of a nature extended to infinity, to the point of surpassing common perceptions or any attempt at ordering, meets the path of perception that we have pursued throughout this study: when the human seeks to perceive nature by approaching it; to the point of merging with it and eradicating the artificial separation between human and nature; to liberate new powers of natural perception decentralized from the human, making the most of Gilles Deleuze's maxim: "There is no art that is not a liberation of the power of life."[28]

Italian philosopher Emanuele Coccia, in *Sensible Life: A Micro-ontology of the Image*, demonstrates the sensible as the being of images, associating the genesis of images with the becoming sensible of things, and which is primarily a force (*physis*).[29] For Coccia, the image is precisely the fruit of this trajectory: "any thing that ends up existing outside of its own place".[30] In the chapter "The Primacy of the Sensible", where he shows how phenomenology has acknowledged the primacy of perception over consciousness but without managing to make the sensible autonomous from the perceiving subject, he formulates in philosophical terms what a filmmaker like Karel Doing accomplishes in action through an artistic procedure: the sensible detached from the eye; when plant beings constitute themselves into images by their own means and constituents; because "it is the existence of the sensible that makes sensation possible and not the contrary: it is indeed because there is the visible that vision is possible".[31] When cinema gives us back the objects of our perception but as subjects themselves perceiving, according to other modalities than ours, we indeed move from perception to the sensible.

These reflections intersect with those of another contemporary French philosopher, Baptiste Morizot, who postulates, with the art historian Estelle Zhong Mengual, that the ecological crisis we are going through results above all from a "crisis of sensibility" and a perception devoid of consideration for the living.[32] Indeed, he observes that the human has rejected under the term "nature" the totality of the living world which was not him, relegated to a simple decor and nevertheless reserve of resources, a place of fulfilment and a symbolic support, depriving the living world of the proper qualities that should define it, and this because we have ceased to pay any appropriate attention to it: "When the living world falls outside the field of collective and political attention, outside the field of what is deemed important, then a crisis of sensibility is triggered."[33] We can then recognize in the effort of attention of experimental filmmakers directed towards what has been abandoned by human fictions the epistemological lever of "tracking" as practised and theorized by Morizot: "Tracking, in the broad sense of an investigative sensitivity towards living beings, is a very clear experience of the way we can access the meanings and communications of other life forms".[34] And, one might add on the basis of the filmic examples studied: their own modes of perception. How then can we adjust our consideration, as the tracker-philosopher suggests, in the face of this mechanized, deanimated and disenchanted living world without which we simply could not remain alive ourselves: by freeing ourselves from regulated mechanics (those of machines as well as those of the standardized forms); by reanimating the living with its own movements; by putting our own sensible powers at the service of images, and by allowing them to gain the status of images; and by inventing new paths for the sensible.

Notes

1. Most of the films studied here were shown in the context of the Preview Show of the distributor Light Cone in Paris in June 2021, which consists of presenting each year the new films entered in the catalogue.
2. Henri Bergson, "The Life and Work of Ravaisson", *The Creative Mind. An Introduction to Metaphysics* [1934], trans. Mabelle L. Andison, New York, Dover Publications, 2007, p. 265 [quoted by P. Hadot].
3. Pierre Hadot, *The Veil of Isis. Essay on the History of the Idea of Nature* [2004], trans. Michael Chase, Cambridge, MA, Harvard University Press, 2006, pp. 211–229.
4. Hermann Hesse, *The Art of Idleness* [1904], trans. Tim Newcomb, Newcomb Livraria Press, 2023, p. 7.

5. "The Structure of Film", in *Film Form. Essays in Film Theory* [1949], ed. and trans. Jay Leida, New York/London, A Harvest/HBJ Book/Harcourt Brace Jovanovich, 1977, p. 172.
6. Hans Blumenberg, *L'Imitation de la nature et autres essais esthétiques*, trans. Isabelle Kalinowski and Marc de Launay, Paris, Hermann Publishers, 2010.
7. Jacques Tassin, *Pour une écologie du sensible*, Paris, Odile Jacob, 2020, pp. 11–12. Unless otherwise noted, all translations from original French are by the author and translator.
8. "In order to *see* things, we must first of all look at them as if they had no meaning, as if they were a riddle. ... To understand less, to be ingenuous, to remain stupefied: these are reactions that may lead us to see more, and to take account of what lies deeper, what is closer to nature. ... It seems to me that defamiliarization may be a useful antidote to the risk we all run of taking reality (ourselves included) for granted. The anti-positivist implications of such an observation are obvious." Carlo Ginzburg, *Wooden Eyes. Nine Reflections on Distance* [1998], trans. Martin Ryle and Kate Soper, New York, Columbia University Press, 2001, pp. 7–22.
9. Augustin Berque, *Descendre des étoiles, monter de la Terre. La trajection de l'architecture*, Bastia, Éditions Éoliennes, 2019.
10. Jacques Tassin, *Pour une écologie du sensible*, *op. cit.*, pp. 28–29.
11. In the words of Ivan Illich, the observation is unequivocal: "Radical monopoly makes people prisoners of welfare. Men overwhelmed by commodities are rendered impotent and in their rage either kill or die", in Ivan Illich, *Tools for Conviviality*, New York, Harper & Row, 1973, p. 60.
12. Gilles Deleuze and Félix Guattari, *A Thousand Plateaus: Capitalism and Schizophrenia* [1980], trans. Brian Massumi, Minneapolis, University of Minneapolis, 1987, p. 481.
13. Bruno Latour, *Down to Earth. Politics in the New Climatic Regime* [2017], trans. Caroline Porter, Cambridge, Polity Press, 2018, p. 83, cited in Jacques Tassin, *Pour une écologie du sensible*, *op. cit.*, p. 39.
14. On the Vimeo page of the filmmaker: https://vimeo.com/carynyc.
15. The Purkinje effect refers to the fact that in humans, colour vision depends on brightness: at high brightness, it is the red colours that the eye distinguishes best but when the illumination decreases, it is the blue colours that are best perceived.

16. Jacques Tassin reminds us that "Current sciences still do not have a model of representation of *lived life* [*la vie vivante*]".
17. "The further we move away from Marey in time, the more his presence insists in the field of experimental images, in the manner of a light that takes years to reach us but shines all the more intensely", Nicole Brenez, "Étienne-Jules Marey: modèles et investissements problématiques dans le cinéma expérimental récent", *in* Dominique de Font-Réaulx, Thierry Lefebvre, Laurent Mannoni (eds), *EJ Marey. Actes du colloque du centenaire*, Paris, Arcada Éditions, 2006, p. 121.
18. We could also mention the very beautiful *Waves* (1973) by Patrick Carey, available online on the website of The Irish Film Institute: https://ifiarchiveplayer.ie/waves/ [consulted on 12 September 2021].
19. É.-J. Marey, *Album de physiologie artistique*, no. 1, "Des mouvements de l'homme", Paris, Société d'études scientifiques, 1893, quoted in Dominique de Font-Réaulx, "Vagues fixes et en mouvement, autour du film *La Vague* d'Étienne-Jules Marey", in D. de Font-Réaulx, Thierry Lefebvre, Laurent Mannoni (eds), *EJ Marey. Actes du colloque du centenaire*, *op. cit*., p. 49.
20. https://www.jacquesperconte.com/oe?242.
21. Joyce Lainé and Jacopo Rasmi, "Laisser infuser les images dans les ruines de l'industrie cinématographique", in Conference "Les laboratoires cinématographiques partagés. De la cave au jardin, formes et fonctions d'un lieu d'invention et de résistance", coordinated by Éric Thouvenel, Rennes, 2 July 2021.
22. "The emphasis on community and process (Process Cinema) at the Film Farm Retreat encourages artists to embrace the unexpected and environmental temporalities through film, plunging into celluloid through different chemical and biological (and 'green') processes". Philip Hoffman, "Vulture Aesthetics: Process Cinema at the Film Farm", in Luisa Greenfield, Deborah S. Phillips, Kerstin Schroedinger, Bjorn Speidel, Philip Widmann (ed.), *Film in the Present Tense: International Symposium on Current Developments in Analog Film Culture*, Berlin, Archive Books, 2018, pp. 37–38.
23. *Ibid*., pp. 39–40.
24. Grégory Quenet, *Qu'est-ce que l'histoire environnementale ?*, Seyssel, Champ Vallon, 2014, p. 15.

25. Karel Doing, "Phytograms: Rebuilding Human–Plant Affiliations", *Animation: an Interdisciplinary Journal*, vol. 15, no. 22–36, 2020, p. 32.
26. Gene Youngblood, *Expanded* Cinema, New York, P. Dutton & Co., 1970, pp. 41–44.
27. Marcel Conche, *Présence de la nature*, Paris, PUF, 2001, p. 39.
28. Pierre-André Boutang and Michel Pamart, *L'Abécédaire de Gilles Deleuze*, 1996 (chapter "R comme Résistance").
29. Emanuele Coccia, *Sensible Life. A Micro-Ontology of the Image* [2010], trans. Scott Alan Stuart, New York, Fordham University Press, 2016, p. 12.
30. *Ibid.*, p. 33.
31. Emanuele Coccia, *La Vie sensible*, trans. Martin Rueff, Paris, Rivages Poche, 2010, p. 60 (NB: this sentence is found in the French version of the text and not the somewhat abridged English version).
32. Baptiste Morizot, *Ways of Being Alive* [2020], trans. Andrew Brown, Cambridge, Polity Press, 2022; as well as the work of Estelle Zhong Mengual, *Apprendre à voir. Le point de vue du vivant*, Arles, Actes Sud, 2021.
33. Baptiste Morizot, *Ways of Being Alive*, *op. cit.*, p. 6.
34. *Ibid.*, p. 103.

CHAPTER 22

From Observation to Techno-natures: Ecological Concerns in Moving Images by Contemporary Artists

Lukas Brasiskis

Towards an Alternative Lens on Environment

Concern over the impact of human activities on the environment in recent decades has gained a new relevance, prompting scientists to pronounce a new geological epoch titled the Anthropocene. The epoch of human domination over the planet marks a number of ecologically disastrous outcomes, the global climate's departure from its natural course being a central issue. The troubling outcomes of the Anthropocene, such as the extinction of plant and animal species, rising sea levels, radioactive contamination, biodiversity loss, air and water pollution, as well as the social repercussions of all the above, have acquired greater significance

L. Brasiskis (✉)
Brooklyn College, Brooklyn, NY, USA
e-mail: lb2892@nyu.edu

Columbia University, New York, NY, USA

E. Della Noce and L. Murari (eds.), *Expanded Nature*, Experimental Film and Artists' Moving Image,
https://doi.org/10.1007/978-3-031-70729-2_22

among moving-image artists too. In the face of the increasing environmental crisis and political impotence to react to it globally, eco-conscious filmmaking has been rethought and re-practised, as contemporary filmmakers and artists across the world continue to voice problems related to the human impact on the Earth's ecosystems and atmosphere. A variety of moving-image works representing the ecological crisis and pointing towards a less anthropocentric, economically and environmentally fairer future for both humans and non-humans have been produced in the past two decades.

Ranging from observational documentaries to fictional stories, creative simulations and archival appropriations, we have been witnessing the increased presence of moving images within the world of contemporary art. As Erika Balsom puts it, the divisions between filmic installations and single-channel films play an important role in thinking about the institutional and economic situatedness of the moving image vis-à-vis contemporary art; however, in this chapter I do not over-emphasize *dispositif*-related differences between single-channel and multi-channel installations because I analyze the eco-conscious representations of nature ranging from projections to installations.[1] Treating artists who employ cinematic aesthetics (what Jean-Christophe Royoux calls "cinema of exhibition"[2]) and film directors who install their films in galleries (what Catherine Fowler refers to as "gallery films"[3]) equally, I focus on the ways the images of nature make it to the screen and what affects they produce.

Observational documentation is the most common artistic strategy for capturing images of the planet facing environmental issues among artists whose works are represented in between the black box and the white cube, to paraphrase the title of Andrew Uroskie's study of expanded cinema.[4] The positioning of the camera in otherwise inaccessible *milieus* with an attempt to convey the experience of wilderness and decentralize the anthropocentric point of view is a common strategy applied by both video artists and filmmakers. A conventional example of the observational representation of violated environments is exemplified by *The Anthropocene Project*, an art project by Nicholas de Pencier, Edward Burtynsky and Jennifer Baichwal.[5] Functioning in the form of an exhibition based on observational documentary and relying on the inside space of the artwork (without much reflection on what Catherine Fowler calls "the outside space of the gallery"), the project relies on recording and exhibiting visually stunning images of disastrous landscapes—all expected to excert a spectacular effect on the spectators. However, the recent spike of highly aestheticized visions of environmental disaster has triggered a discussion on the limits of this kind of "Anthropocene visuality". Given the unclear outcomes of observation, one should raise a question: what do spectacular images of a devastated nature tell us about the actual problems of a world facing environmental crisis?

The causes of environmental problems lie in particular forms of human activities: capitalist, colonial, consumerist and patriarchal.[6] Therefore, the environmental crisis is not just a consequence of abstract biological processes or the result of our lack of contemplation of nature, but the outcome of an exploitative, extractive and polluting administration of the world. The Anthropocene is characteristic of the complex layers of interwoven scales, as the art historian T. J. Demos emphasizes. Having this in mind, any attempt to visualize the Anthropocene seems to be far from objective—images of nature in crisis do express a position of ideology, economy and power—and so the crisis cannot be laid bare as yet another spectacle. According to media and culture theorist Nicholas Mirzoeff, conventional Anthropocene visuality "frequently acts as a mechanism of universalization ... inadvertently making us all complicit in its destructive project".[7] Therefore, as Demos and Mirzoeff emphasize, the abstract and spectacular visualization of human-caused ecological problems works ideologically in support of a capitalization of nature, of anthropocentric political economies, as well as being an endorsement of geo- and bio-engineering. And so here we encounter a contemporary paradox: to visualize the current environmental crisis properly, the moving-image artist has to renounce the idea of imaging nature and natural sites as if they existed outside of history, independently from political economy and human-made technology. In other words, one has to focus on the entanglements between the manufactured and the organic, giving up the clear-cut distinction between enlightened culture and untamed (and to-be-saved) nature.

New materialist and eco-critical thought challenging the ontological and epistemological divide between nature and culture as well as human and non-human has been widely applied in art theory and criticism with the aim of reconsidering and expanding the analysis of art (including moving images) in the time of the Anthropocene. "Natureculture", a conceptual term introduced by the feminist scholar Donna Haraway in her *The Companion Species Manifesto*,[8] is commonly used as one of the ways to overcome the centrality of the human and the purity of nature in fathoming the human–nature relationship anew. This term perfectly illustrates what is at stake in thinking about (and visualizing) nature in the Anthropocene. Dissolving the modernist borderline between the natural and the cultural, "natureculture" marks a refusal to treat nature either as a pristine object for human gaze or as a useful resource for domestication, administration and use. This attitude does not suggest that portrayals of forms of natural life have ceased to be valuable subjects for artists' films; it rather suggests—as some works I pinpoint below can prove—that non-human agents cannot be naively singled out. Instead, they should be treated inseparably from their entanglement with human-made technology.

"Techno-nature", a derivation from "natureculture", the notion I apply to discuss recent artists' films, exposes complex entanglements between technology and natural forces. Posing an inseparability between the artificial and the organic, the term helps to fathom new ways environments are depicted in time of the Anthropocene and prompts anti-essentialist readings of images of the technology–nature relationship in moving-image works by contemporary artists. In what follows, I review a few common themes among artists' films and installations that strive to visualize techno-natures comprehensively, combining local, regional and global perspectives. Using the notion of techno-nature as a point of departure, I focus on works that do not mirror nature for humans, but expose a constantly developing techno-economic reality, of which nature is an inextricable part. Through visualization of the causes for and the solutions to the pending environmental crisis, the works discussed in this text thus visualize the Anthropocene and its infrastructural premises as a complex system of interwoven human and non-human agents and focus on the interrelations between natural and cultural forces that have become more important than ever.

Forays into the Underground: Visible and Invisible, Fictional and Real

The recognition of mankind as a major environmental force goes hand in hand with the understanding of a wide scale of industrial and corporative extraction of natural resources. Extraction and excavation in all of their forms—from mining and drilling to exploitation of soil, water, plants and other resources—have become a general theme in audiovisual works that attempt to visualize techno-natures at work (Fig. 22.1).

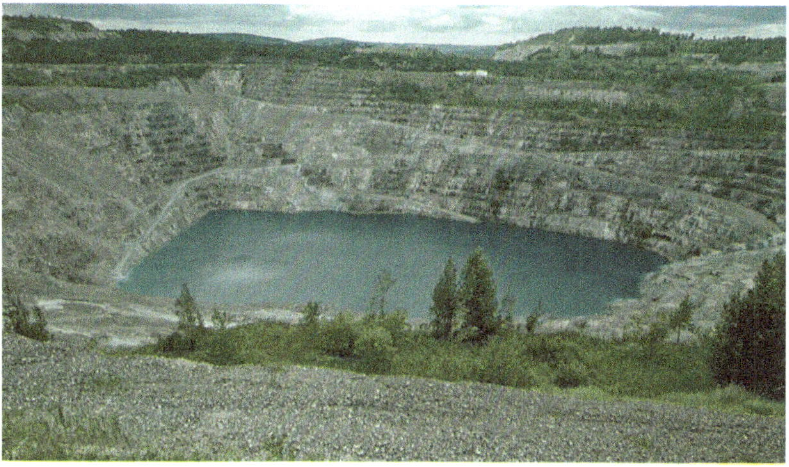

Fig. 22.1 Sasha Litvintseva and Graeme Arnfield, *Asbestos* (2016) © All rights reserved by the artist/Courtesy of Light Cone

A number of moving-image artists, David Kelley and Patty Chang (*Spiritual Myopia*, 2018), Sasha Litvintseva and Graeme Arnfield (*Asbestos*, 2016), Thirza Cuthand (*Extractions,* 2019), Zhao Liang (*Behemoth,* 2015) and Ben Russell (*Good Luck,* 2017), to mention just a few, have chosen the extractive industry as a subject matter for their films and videos in the search for alternative methods of rendering ecological concerns into images. Responding to the similar set of concerns, *Mineral Crafts* (2018), a five-channel video by the Taiwanese artist Chia Wei Hsu, scrutinizes the history of the extraction of copper. The installation is based on computer-generated images simulating Jinguashi, a town in the Ruifang district in Taiwan that is known for its gold and copper mines. Virtually connecting the geographical features of Jinguashi's landscape and the development of the mining industry during the times of Japanese occupation, *Mineral Crafts* tells the story of how the extraction of natural resources has impacted human lives on a massive scale. Highlighting the key role that copper mining plays in humanity's past and present, the work by Wei Hsu focusses on daily applications of copper (such as the utilization of copper for the production of household objects, military equipment and electronics) and wittily yet indirectly juxtaposes the human needs that propel the mining of natural resources with the ecological problems that mining has engendered. Style-wise, *Mineral Crafts* mimics *Minecraft*, a widely known video game that tasks players to colonize the empty virtual landscapes through extracting raw materials, crafting tools and items, and building structures. In collaboration with the gaming community, the artist designed the animated image of the *Minecraft*-like landscape of Jinguashi featuring natural sites, the copper refinery and the Japanese-style residence building. A voice-over for the animated film, recorded using Google's Text-to-Speech software, is reminiscent of virtual guides commonly used to introduce gamers to the historical context a particular video game is based on. A didactic explanation of how copper mining propelled human society from ancient times to the Information Age and how it created the material conditions for the existence of *Minecraft* and the entire digital realm performs a double function. On the one hand, it makes a visualized genealogy of materiality (from physical things to "internet of things") easily fathomable. On the other hand, it functions as a satirical take on the ideology behind anthropogenic diggings into the Earth's strata ("This is where you humans and I could meet, mine the resources and build our dreams", says the narrator/character ironically within the diegetic world of the game-like animation). Take the beginning of *Mineral Crafts* when the guide explains: "if there is a mistake, no worries, just knock it out with the left click!" The corporate-sounding voice-over leads one to think about the impossibility of such an "undo" option in our current reality where the planet faces the threats of the Anthropocene. The spatial installation of the work matters as well. The main screen divided into three parts is accompanied by two monitors representing the environments of *Minecraft* that mimic apparatuses used to play video games. The tension between the screen and the monitors puts an emphasis on the technological nature of the video game itself. All in all, using the strategy of expanded cinema, Chia Wei Hsu expands the context of copper mining from the single region to the global history of the development of techno-natures that includes the artwork itself as well as the video game to which the animation makes reference.

Lithium Lake and the Lonely Island of Polyphony (2020) is another study of excavation and distribution of natural resources. A three-channel video installation by emerging Chinese moving-image artist Liu Chuang is characteristic of a complex audiovisual examination of the global economy in *longue durée* with a special emphasis on the extraction of lithium—the lightest and most highly reactive metal on the planet. A visual homage to the eponymous Stanley Kubrick's *2001, A Space Odyssey* (1968) (referred to, among some other noted sci-fi texts, such as Liu Cixin's novel *The Three-Body Problem* (2008) and James Cameron's movie *Avatar* (2009)) serves as a framework for the story about technology's role in human history that is told in the future perfect tense from (apparently) aliens' perspective. As the work makes clear, lithium ions—the core component of carbon-free power supplies—are in increasingly high demand. They are refined from deposits in places such as the Andes Mountains in Chile and the Tibetan Plateau in China. Liu Chuang weaves images from these excavation sites with aural examination of the history of polyphonic music, creating a multilayered exploration of the interconnections between technology, nature and human. The artist uses music theory to highlight the difference between monophonic and polyphonic sounds as an allusion to prototypical technology, which not only mirrors the form of the three-channel film itself (the work's narrative structure is based on different threads connected alongside and within the three screens), but also functions as an audiovisual metaphor that helps to envision a speculative alternative to the industrialized and extractive world of multinational corporations. As the three-channel video suggests, the functioning of polyphony, which is based on connections of each single sound to the entire network of all sounds, could be applied as a model for "unlocking" technology from its exploitative use in the mining industry and moving towards a more sustainable carbon-free world. In times of rising concerns regarding the role China's extractive activities play in the global crisis of the Anthropocene, Chuang's astoundingly audiovisual and yet ingeniously speculative work demonstrates the original way to respond to the environmental malfunctions (*sic*) of the current neoliberal economy.

How to react to human-caused disasters beyond a documentary portrayal of environmental harm? Can one expose ecological problems without claiming the authenticity of the footage? The questions raised by Chia Wei Hsu and Liu Chuang are also tackled by other artists. An erasure of boundaries between real and simulated environments resulting in abstract yet critical takes on humans' administration of nature is a conceptual thread connecting the moving-image installations by British video artist John Gerrard. *World Flag* (2017), a video installation by Gerrard that features glimpses of the Amazon, Nile and Yangtze, among other multi-continental rivers, is characterized by close-ups of opalescent, aesthetic shapes floating on surfaces of water. The title of the work is a linguistic metaphor of, and an obvious reference to, the global pollution of the world's waters caused by humans' industrial activities. The shapes of shimmering colours that appear to be gasoline and oil spills covering the water were in reality digitally simulated and programmed to alter according to the time of the day one visits the installation. Gerrard's faux hyper-realist video invites the audience to believe that what we see are real marks of widespread water pollution, despite the fact that it is actually a digital simulation. Based on a meaningful digital reenactment of truth, *World Flag* exemplifies the technological power of superimposition of human infrastructures and natural sites. Through its faux digital gesture, the film visualizes the conditions of the Anthropocene, where nature cannot be considered separately from the technology with which the image-making itself is entangled.

Another hypothetical work mixing fiction and facts is *The Disquiet* (2013), a film by Lebanese artist Ali Cherri. The single-channel film investigates the geological situation in Lebanon, a country situated on the site of several major geological faults. Looking for the traces of impending disaster, the single-channel film links geological and seismological events to the sociopolitical history of Lebanon, emphasizing wars, political upheavals and a number of social revolts that the Lebanese have faced in recent decades. In his exploration of Lebanon's intense history, Cherri looks at the interdependence between its seismic conditions and numerous human-made conflicts. Through seamless editing together of natural and political catastrophes, the artist invites the spectator to consider longer timescales than those of recent historical events. As Cherri has pointed out in his master class at Harvard University, research for the film began from an interest in how and why disasters, including natural catastrophes and human-instigated wars, have become the indefinite state of things in Lebanon.[9] By reflecting on experiences of slow and non-spectacular violence, the film shifts its attention from the underground to the surface, from the invisible to the visible. The artist edits the archival footage of the historical events together with the footage of positivist scientific data of seismic activities. In the artist's own words, "even with the use of a witness, or a text, or a photograph: catastrophe is always off-screen, beyond our grasp. With the long shots of seismometers registering on paper or on monitors the movements of tectonic plates, I wanted to highlight our position as witnesses; as we observe the catastrophe in the making."[10] The very beginning of *The Disquiet* is reminiscent of an observational documentary; however, paradoxically, it slowly gravitates towards techno-scientific fiction (Fig. 22.2).

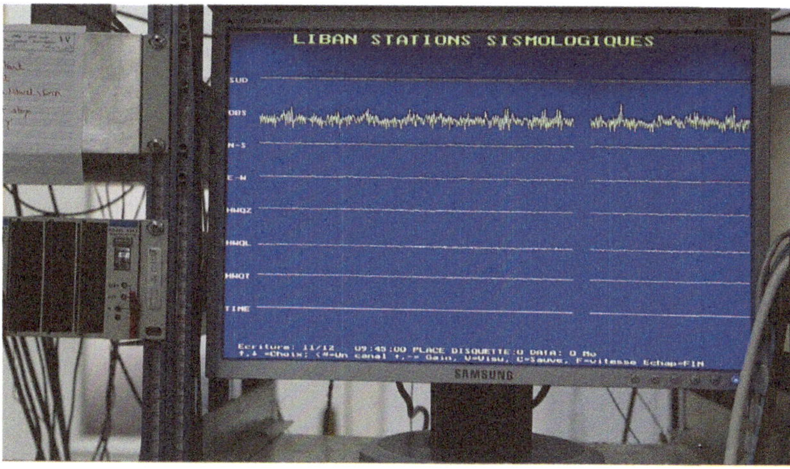

Fig. 22.2 Ali Cherri, *The Disquiet* (2013) © All rights reserved by the artist/ Courtesy of Light Cone

By drawing attention to positivist attempts to image and explain the natural catastrophe, Cherri reshapes the common view of knowledge about the history of the region. The traces of the imminent destruction become visible due to the focus on the entanglement of natural forces and techno-scientific knowledge. The film illustrates that the images of anthropocentric truths always reveal themselves as something new, even if that something has been there all along. In other words, as the artist implies, natural disaster is always present and inextricable from history, even if it does not show up in our mundane lives and thus cannot be processed due to the limits of our habitual non-scientific perception. The complex dynamics between visible and invisible thus pertains to ethical and political choices related to our stance towards history. *The Disquiet*, among the other works mentioned above, deconstructs the romantic idea of a pure and pleasing nature waiting to be saved during a time of crisis. Instead, it invites the spectator to reconsider the dichotomies between invisible and visible, true and simulated, pure and artificial, and ponder over the importance of approaching the bleak effect of human ambitions by intentionally mixing real and fictional entanglements of nature and technology.

Decolonization of Nature and Critique of Modernization

According to Demos, it is not universal humanity but corporate economy that is responsible for the current environmental disturbance. The ecological crisis should therefore be considered technocratic and so all the attempts to create a counter-visuality to the stereotypical Anthropocene visuality should start with an invitation to "decolonize nature".[11] Jason W. Moore, in his book *Anthropocene or Capitalocene: Nature, History and the Crisis of Capitalism* (2016), adds that "capitalism is a way of organizing nature as a whole … a nature in which human organizations (classes, empires, markets, etc.) not only make environments, but are simultaneously made by the historical flux and flow of the web of life. In this regard, capitalism is a world-ecology that joins the accumulation of capital, the pursuit of power and the co-production of nature in successive historical configurations".[12] Thus, Moore refers to the corporate economic system as a specific agent which not only consumes and exhausts natural resources but also manages to overlook the consequences for the poorest and most underprivileged populations. How does one interrupt the narrative of modernist progress that glorifies administration and colonization of nature through the process of technologization?

The *First Man Who Bought a Juicer Bought It Not for Drinking Juice* (2008), Lu Chunsheng's short film, presents a farmer from Texas and a reaper machine, the two main characters of a film that tells a non-climactic story of the repetitive acts of both the machine and its human owner, casting a relationship in which humanity (as well as nature, hence the dead animals portrayed around) is doomed to be enslaved by the technology it has created to serve itself. The interaction between the two makes up a satirical portrayal of a modern-day human–machine assemblage. Appropriating genre tropes common to both slow documentary and science fiction, the wordless story portrays the man repairing and advancing the mechanical machine before the technological creation undermines everyone and enslaves its own creator. The automated machine-like actions of the farmer represent the consequences of the teleological state of being brought about by the industrial revolution. Lu Chunsheng's film diametrically reverses the all-too-familiar narrative of progressive modernization and technological modernization and provides a subtle critique of the industrial logic applied in the process of colonization of nature.

The drive to decolonize the rigid anthropocentric structures of the past and the present is also characteristic of the works of the Portuguese moving-image artist Pedro Neves Marques. *Exterminator Seed* (2017) and *The Pudic Relation Between Machine and Plant* (2016), two single-channel moving-image installations by Marques, aim at palpating the hybridity of techno-natural assemblages. *Exterminator Seed* tells a story of an offshore oil rig worker, who, due to an oil spill that has contaminated the Brazilian coast, is evacuated to Rio de Janeiro, where the locals remain ignorant of the incoming disaster. In the city, he meets indigenous android Ywy who convinces him to travel to Mato Grosso do Sul, her state of origin. The worker, being human, is incapable of understanding the android's anti-capitalist perspective favouring interspecies relationships to technological progress. The film was shot on actual locations—soy farms and rural areas in Brazil. Employing long takes and accompanied by ambient sounds, it defamiliarizes the image of human-led industry by providing a screen agency to the environment this industry exploits. The worker cannot understand a manifesto-like monologue about reproductive rights that Ywy delivers; only non-human plants are capable of hearing and understanding her points.

Machinic Bodies, Hybrid Natures and Non-binary Voices

The Pudic Relation Between Machine and Plant, another work by Pedro Neves Marques, is a fictional video that continues a speculative exploration *of techno-natures.* The video portrays a disembodied robotic arm repetitively caressing Mimosa Pudica, a plant that characteristically closes in on itself when touched. Through this formal experiment, the work alludes to material practices in the post-human world as a denunciation of phallocentrism. By subtly referencing environmental colonialism, *The Pudic Relation Between Machine and Plant* invites us to reconsider the traditional understanding of sexuality, technology and science. It not only illustrates that the decolonization of nature can be based on a critique of modernization, but it also implies a decolonization of the look that helps to highlight the suppressed non-binary angles of the world and expose hybrid bodies and voices. Influenced by science fiction and speculative storytelling, Marques offers a subtle yet politically charged vision of a new type of coexistence between humans and non-humans beyond the binary conventions of reproduction and fertility (Fig. 22.3).

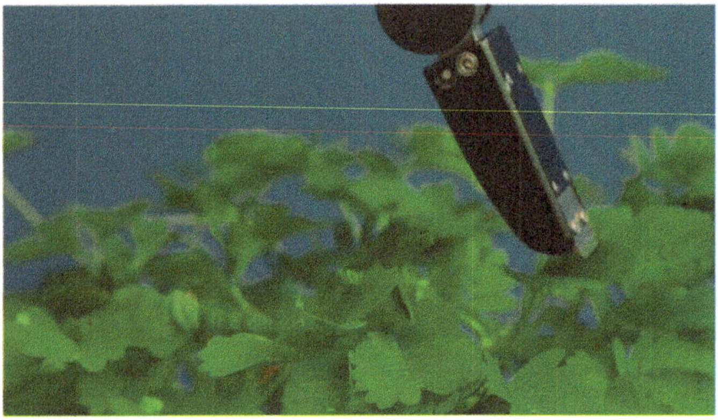

Fig. 22.3 Pedro Neves Marques, *The Pudic Relation Between Machine and Plant* (2016) © All rights reserved by the artist/Courtesy of Light Cone

The Lithuanian artist Emilija Škarnulytė in her single-channel installation *Sirenomelia* (2017) continues these explorations of the limits of anthropocentrism through the rejection of distinctions relying on binary logic. Škarnulytė's work was shot in Olavsvern, Norway within a 25,000m^2 NATO submarine base that ceased to function after the end of the Cold War and was permanently closed in 2002. The work portrays an underwater journey of a mermaid performed by the artist herself. Incarnated by the artist, the mermaid encounters colourful sea beds, drowned vessels and industrial ruins. The viewer is introduced to the gigantic infrastructure of the ex-military base and is entangled into a subaquatic world of underwater tunnels full of constructions and radars, now populated with various kinds of sea organisms. The image of the mermaid, a quasi-human creature, has persisted as a feature of popular myths and cultural expressions—both entertaining and moralizing—for centuries. A post-modern interest in mermaids is, however, less often imagined in regard to ancient mythical stories, and more often in regard to sociopolitical and identity issues. Contemporary feminist theory regards the mermaid as a figure of empowerment, reconsidering its function from the mysterious and magical to the monstrous, emphasizing its ability to blur gender and social dichotomies. Rosi Braidotti—to mention just one of many feminist thinkers—explains the monster's/mermaid's ability to simultaneously secure and destabilize our perceptions of selfhood. The monster/

mermaid, in Braidotti's words, "is neither a total stranger nor completely familiar; s/he exists in an in-between zone [and] helps us understand the paradox of 'difference'...".[13] Škarnulytė describes the mermaid in her video in a surprisingly similar way:

> She is not a stereotypical Mermaid. She is hairless, a woman-torpedo, or perhaps not necessarily a woman, perhaps transgender, but more a new species that have adapted to live in different conditions when there are possibly no people left. She is lonely and sensitive. The space around her is gigantic, aggressive and masculine, the same as in most projects: CERN, mines, underwater stations, and the world of astronomy. This mermaid reclaims the ocean in a nonviolent way or rather shows a dimension of the ocean that cannot be appropriated by war.[14]

"She/they is a cyborg," the artist says in another interview, "linked to a human just merged with the fish, submarine, machine...."[15] Thus, the character of the mermaid helps the artist question traditional binary social and gender norms and power structures. Spatially immersive, Škarnulytė's work is—I suggest—inseparably related to the posthumanist performativity that helps overcome the system of binary dualisms. Due to an angled screen, the spectator acquires an affective POV of the sea creature, as one would flow through in place of the mermaid. "What if the political generativity [of films] actually lies in learning to listen to the myriad voices of the world in which we find ourselves—today, now—voices from which we can no longer (if we want to continue to survive on this planet) claim to separate ourselves based on the facile assertion that we have language, or speech, and they do not?", ask Eduardo Kohn and Lisa Stevenson.[16] The figure of the mermaid in the film is thus an utterly liminal techno-natural creature that confounds the hegemonic boundaries of gender and species and subverts normativity in a variety of ways, from challenging traditional masculine and militarist perceptions of the Cold War to expanding how one understands and communicates being in the world vis-à-vis the material remnants of the Anthropocene.

O peixe (2016), a short single-channel film by the Brazilian artist Jonathas de Andrade, shot in collaboration with fishermen from the towns of Piaçabuçu and Coruripe at the mouth of the São Francisco River, presents yet another vision of surpassing the binary understanding of human–nature relationships. In a series of visually astounding vignettes, the film portrays human–nature assemblages that do not have much visibility or voice in art. De Andrade's camera observes individual fishermen on their boats as they catch and then hold the fish to their chests before they pass away, staging for the camera the act of a performative ritual that manifests in an amiable care for their prey. The film is intentionally provocative. On the surface, the intimate carrying of the fish reminds us of an erotic gesture, and yet one is also exposed to the anthropocentrism that pervades the film. In its attentive style, *O peixe* is reminiscent of ethnographic documentaries, although the absence of dialogue generates an ambiguity of meanings and triggers slowly altering reactions on the spectator's part: from grief caused by the exposure to the process of dying, to an emotional experience of fake solidarity between human and fish, to the erotic pleasure of witnessing the animistic ritual. All in all, the film reveals a natural asymmetry between humans and other life forms by non-judgementally pointing to the human violence as benevolence and as necessity. As de Andrade himself notes:

> The fish is about the collapse of the relationship between man and nature, but also about working, surviving. And it's about a village that lives in harmony with the environment, asking permission. Nonetheless, it is a fatal hug, one of domination. In terms of the Anthropocene, it becomes a metaphor for the total failure to limit exploitation, the normalization of killing, unbounded consumerism, self-serving environmental devastation. But the image could also be read as a simple idea of surviving on this planet.[17]

In Lieu of Conclusion: New Frontiers

As the cases described above exemplify, artists in the world of ecological crisis can be seen as pioneers exploring the new frontiers of the possible representations of techno-natures. The creative attempt to construct dialogues with interdisciplinary researchers, scientists and practitioners is a common thread among new generations of moving-image makers. Their works bear characteristics of a socio-economic, ecological and cultural laboratory striving to visualize the more-than-human that was taken to be invisible for too long. In their search for ways of doing this, the artists' films not only visualize nature, but also focus on unheard voices and invisible communities, and challenge the image of nature through the lenses of postcolonial and feminist thought. Atmospheric soundscapes, unconventional camera work and unconventional (sometimes multi-channel) editing open doors into speculative worlds where care, desire and exploitation of the non-human are being rethought vis-à-vis techno-natural assemblages.

Moreover, the works discussed illustrate that in the age of advanced technology, the images of nature and environment, be they of non- or post-human creatures, the Earth's strata or human-built infrastructures, should be read as if they were a script that needs to be interpreted, a trace of its own past inseparable from human activities. Trevor Paglen's *Untitled (Reaper Drone)* (2015), a single photograph of a miniature drone the artist captured against a luminous Nevada morning sky, can function as an overarching example of the search for the alternative visualization of techno-natures. As the artist reveals, the nearly imperceptible figure of the drone was operated from a remote US military base. However, due to the distance from the scene, the spectator cannot see the drone and instead of getting straightforward information feels immersed in the oversaturated sky that evokes both Romantic landscape paintings and modernist history of abstraction. Paglen thus blurs the line between art and investigative journalism, focussing on what cannot be clearly represented—the secret military site—and so reveals the connection between nature and contemporary warfare. As a result, nature appears as an effect of military, economic and political systems. *Untitled (Reaper Drone)* perfectly exemplifies the current state of ecological concerns among contemporary moving-image artists who are devoted to artistic practices of image-making that take into consideration the post-natural and hybrid condition of the world in crisis. Paglen's photographic work interrupts the narrative of modernist progress through its stress on capitalist techno-logic that governs the functioning of the Anthropocene and that has become the subject matter for many recent artists' films.

Notes

1. Erika Balsom, *After Uniqueness. A History of Film and Video Art in Circulation*, New York, Columbia University Press, 2017, p. 3.
2. Jean-Christophe Royoux, "Cinéma d'exposition: L'espacement de la durée", *Art Press*, no. 262, 2000, pp. 36–41.
3. Catherine Fowler, "Room for Experiment: Gallery Films and Vertical Time. From Maya Deren to Eija Liisa Ahtila", *Screen*, vol. 45, no. 4, 2004, pp. 324–344.

4. Andrew V. Uroskie, *Between the Black Box and the White Cube, Expanded Cinema and Postwar Art*, Chicago, University of Chicago Press, 2014.
5. On September 28, 2018, the Art Gallery of Ontario in Toronto and the National Gallery of Canada in Ottawa presented simultaneous and complementary exhibitions of the Anthropocene project, in conjunction with the Fondazione MAST, Bologna. The exhibitions feature traditional photographic prints, high-resolution wall paintings with film extensions, film installations and augmented reality installations. See the project website: https://theanthropocene.org [accessed May 11, 2021].
6. T.J. Demos, *Against the Anthropocene: Visual Culture and Environment Today*, London, Sternberg Press, 2017, p. 55.
7. Nicholas Mirzoeff, "Visualizing the Anthropocene", *Public Culture*, vol. 26, no. 2, 2014, p. 214.
8. Donna Haraway, *The Companion Species Manifesto: Dogs, People and Significant Otherness*, Chicago, The University of Chicago Press, 2003.
9. Jason W. Moore (ed.), *Anthropocene or Capitalocene: Nature, History and the Crisis of Capitalism*, Oakland, PM Press, 2016, p. 7. More here: https://filmstudycenter.fas.harvard.edu/event/ali-cherri-master-class.
10. Sheyma Buali, "Image and Imagination. Ali Cherri in conversation with Sheyma Buali", *Ibraaz*, platform 006, 6 November 2013, pp. 13–14.
11. T.J. Demos, *Decolonizing Nature: Contemporary Art and the Politics of Ecology*, Berlin, Sternberg Press, 2016, pp. 5–7.
12. J.W. Moore (ed.). *Anthropocene or Capitalocene: Nature, History, and the Crisis of Capitalism*, *op. cit.*, p. 7.
13. Rosi Braidotti, "Signs of Wonder and Traces of Doubt: On Teratology and Embodied Differences", in Nina Lykke and Rosi Braidotti (eds), *Between Monsters, Goddesses and Cyborgs*, Zed Books, London, 1996, p. 135.
14. Agnė Sadauskaitė, "Invisible Structures. The cut of a stratosphere as a reflection of human values. An interview with the filmmaker Emilija Škarnulytė", *Echo Gone Wrong*, June 11, 2018, available online at: http://echogonewrong.com/invisible-structures-cut-stratosphere-reflection-human-values-interview-filmmaker-emilija-skarnulyte.

15. Samman Nadim, "Sirenomelia. An interview with the filmmaker Emilija Škarnulytė", *Vdrome*, December 2, 2018), available online at: http://www.vdrome.org/emilija-skarnulyte/.
16. Lisa Stevenson and Eduardo Kohn, "Leviathan: An Ethnographic Dream," *Visual Anthropology Review*, vol. 31, no. 1, 2015, p. 52.
17. Vivian Gandelsman and Germano Dushá, "In Recife, Jonathas de Andrade takes the world to task for its impending ecological collapse", available online at: https://www.artbasel.com/news/jonathas-de-andrade-profile-ecological-collapse.

CHAPTER 23

Re-enchantments of French Contemporary Expanded Cinema

Charlie Hewison

"Re-enchant!" Such seems to be the new ecopolitical rallying cry in the face of the current disastrous climate situation. Depending on the writer, what needs to be re-enchanted can slightly vary—the world, science, urban spaces, art—but each time, what is expressed is that to effectively tackle the problem, we must first change our perspective and transform our relationships in order to extract ourselves from a certain disenchantment that is singled out as being the primary cause. Meanwhile, in the world of artist-run film laboratories—those crucial sites for the continuation of photochemical experimental moving-image practices—we find similar calls for re-enchantment. This apparent coincidence can inform us both about the specificity of contemporary practices of experimental cinema in France, and allow us to clarify what is meant by this necessity for ecological re-enchantment.

C. Hewison (✉)
Université Paris Cité, Paris, France
e-mail: charlie.hewison@gmail.com

E. Della Noce and L. Murari (eds.), *Expanded Nature*, Experimental Film and Artists' Moving Image,
https://doi.org/10.1007/978-3-031-70729-2_23

The artist-run film lab movement emerged in Europe, particularly in France, in the second half of the 1990s. In 1990, three fine arts students in Arnhem (Netherlands)—Karel Doing, Saskia Fransen and Djana Mileta—salvaged Super 8 equipment and materials that their university was throwing out, allowing them to create Studio één. This first artist-run lab was quite quickly successful and inspired several similar projects, including most notably the Atelier MTK in Grenoble, co-founded by Christophe Auger and Xavier Quérel (members of the Cellule d'Intervention Métamkine), quickly joined by Étienne Caire and Gaëlle Rouard, and which opened its doors to the public by the end of 1992. Very quickly, many filmmakers, not only from Grenoble, but also from Paris, Nantes, Switzerland, Italy and Belgium, begin to use this lab. In July 1995, however, the MTK collective, overwhelmed with requests, announced its intention to "work towards the opening of other labs in other cities".[1] In the subsequent years, artist-run photochemical laboratories were established all over France: L'Abominable in Asnières on the outskirts of Paris, Burstscratch in Strasbourg, Mire in Nantes, XHX in Marseille, etc. To stay in touch, share information, news and advice within this community of "labs", members collectively and semi-regularly published the "independent nitrate gazette" *L'ébouillanté*. In the fifteenth and final issue of this gazette, we find, sandwiched between a catalogue of the latest films produced in each lab and diagrams of projection equipment, an article by Blick, a filmmaker, artist and musician. In it, he attempts to rewrite cinema history by pinpointing its birth at the invention of the Optical Theatre by Émile Reynaud.

As Blick points out, this machine allowed Reynaud to adapt his spectacle "according to the reactions of the public. He could slow down to prolong the suspense, speed up at certain moments, or rewind to admire a certain effect."[2] He uses the apparatus in his *Pantomimes Lumineuses* at the Musée Grévin from 1892. Blick's text ends with this plea:

> Let it be known, a carnivalesque *joie de vivre* accompanied the first steps of screen-images. It is up to us to remember this in order to create beauties that re-enchant us. Let the era of erroneous myths cease! Let us trust no longer thieves and other peddlers of trinkets. Let us craft [*Artisanons*]...[3]

To claim Reynaud as the "progenitor" of cinema is to situate its birth in an immersive, interactive and performed audiovisual spectacle. In this sense, the Optical Theatre of 1892 already corresponds to Dominique

Noguez's definition of "expanded" cinema: "any spectacle that exceeds or modifies in one way or another the cinematographic ritual strictly defined as the projection on a screen, in front of seated spectators, of an image obtained by the scrolling of a film strip in a projector".[4] Cinema was therefore born as an expanded and artisanal artform, and Blick sees in the film lab movement a way to return to this state in order to produce works that can *enchant*.

Similarly, philosophers like Akeel Bilgrami[5] or Silvia Federici[6] advocate a return to an *enchanted* relationship with the world. According to them, the development of capitalist societies induced a loss of awareness of the profound continuity between us, humans, and the multiplicity of other existents—animals, materials, living and non-human—with whom we share this space. This loss of consciousness is the cause both of the looming climate disaster and of the alienating relationships that define contemporary society. The *re-enchantment* called for by Bilgrami and Federici should be understood as a reconfiguration of our social relationships—by re-actualizing *commons*, for example—in ways which can lead us to rediscover the feeling of being part of the world that surrounds us.

In both cases, cinematographic and ecopolitical, aspects of a pre-industrial era are revisited and given new significance: that of the *magic* of the first artisanal audiovisual spectacles or that of the *commons*. Nicolas Rey, co-founder of the l'Abominable laboratory, explicitly associates the artist-run film lab movement with the beginnings of cinema,[7] and Noélie Martin, inspired by the works of Bernard Stiegler,[8] has shown how these labs could be perceived as places where a desire to live differently, collectively, manifests itself through the creation of new film production environments:

> The individuals who have thus found themselves in [these laboratories] ... have followed a same path, which can be considered at once as radical, extreme or utopian. What has, it seems to us, really brought them together is indeed the exercise of an escape, whether it be physical or mental, from a disjointed, divided life.[9]

Thus emerges in both fields a concept of *re-enchantment* as an articulation between the desire to act, create or work differently, and a certain *expanded perception* that would therefore result. Using a few examples of works by the Nominoë collective (Emmanuel Lefrant, Nicolas Berthelot and Stéphane Courcy di Rosa), Silvi Simon and Gaëlle Rouard, we will thereby define an *ecopoetics of re-enchantment*[10] in contemporary French expanded cinema, linking the *enchanted* sensory experiences proposed by these works to the non-alienated, artisanal and collective modes of creation that produce them.

Silvi Simon: The Affected Medium

Silvi Simon is an artist and filmmaker from Strasbourg, co-founder of the Burstscratch association in 1994 with Laurence Barbier and Laurent Berger. After initially developing their films in baths at each other's homes, Burstscratch opened a lab of the same name within Molodoï, an independent art space in Strasbourg. Shortly after the founding of Burstscratch, Silvi Simon moved to Brussels and founded a small artist-run lab in a squat called Rossini, before returning a few years later to Strasbourg and Burstscratch. The filmmaker notes that at that time the labs were integrated into larger collectives and squats, attempts to live and work together in autonomous, collaborative and supportive ways.[11] No one was salaried, everyone shared work and production equipment—the developing spirals, the processing chemicals—which the members tirelessly sought in flea markets and second-hand shops. Apart from a few collective performances, everyone worked individually on their own works. However, working together in the same space had the effect, according to Silvi Simon, of mutually motivating each other to create and of integrating outside perspectives in the elaboration of her work. Silvi Simon's memories of the years at Burstscratch and the squats thus seem to depict an atmosphere of solidarity and *conviviality*, in the sense defined by Ivan Illich, who writes that a "convivial" society is one in which "modern technologies serve politically interrelated individuals rather than managers".[12]

This conviviality is manifested in Silvi Simon's recent works in two ways: an enthusiasm expressed in the manipulation of the cinematographic apparatus and a generosity towards the spectators, invited to be *enchanted* by what this apparatus can do. Indeed, Simon's works always produce a sensation, described by the neo-materialist philosopher Jane Bennett, of being "simultaneously transfixed in wonder and transported by sense", of having both a "pleasurable feeling of being charmed by the novel and as yet unprocessed encounter" and of being "disrupted or torn out of one's default sensory-psychic-intellectual disposition[13]".

In *Dys-Focus N° 1: Nature Collection* (2018), a series of digital installations produced by the artist, this enchantment is linked to interrogating the way in which the medium—the material existence of which is here foregrounded—confronts the world. A first installation in this series, *Prendre le vent* [*Catching the Wind*], projects a view of trees in a high wind in the centre of a screen, and it is the entire image that is agitated; it detaches from the black background of the screen and only the presence of a central fixed branch prevents it from completely flying away. For *Persistance – Écrans phosphorescents* [*Persistence – Phosphorescent Screens*], the filmmaker uses a laser to illuminate point by point the surrounding nature at night in the Vosges mountains. In doing so, she discovers that the laser light is too bright to directly reveal what it points at, but that each physical element with which the laser comes into contact, each blade of leaf etc., deflects the light onto its surroundings and illuminates them: the natural materiality ***makes visible*** that which surrounds it (Fig. 23.1).

Once these images were recorded, the artist created digital overlays, allowing a gradual revelation of the environment in the image, which is then projected onto a phosphorescent screen made of organic material. This phosphorescent material "remembers" the successive illuminations; thus, even when the projection stops, the image of the environment

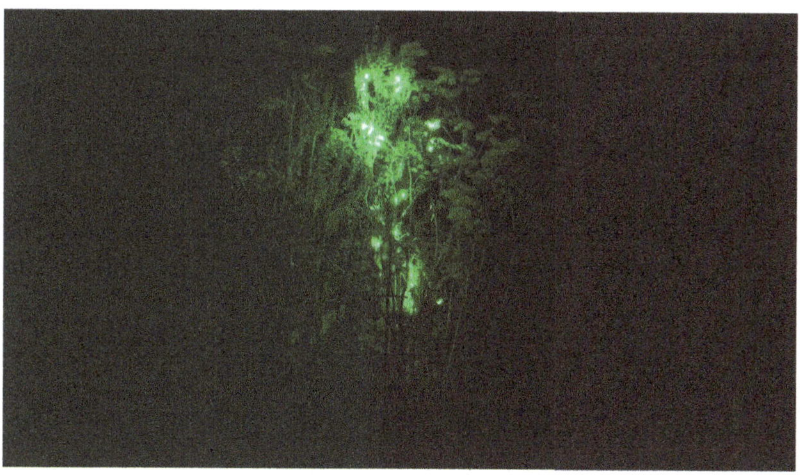

Fig. 23.1 Silvi Simon, *Persistance—Écrans phosphorescents* (2018)

persists, phantomatically, on the screen. Finally, the last installation in the series, *Assemblages Natnumériques* [*Natdigital Assemblages*], consists of several digital screens (mobile phones or tablets) that broadcast shots of a rough sea in low definition, establishing doubt about what we see: pixels or fragments of the environmental elements?

In these installations, Silvi Simon interrogates the ways in which "Nature" affects the very mediums that make it visible to us. The space of the *white cube* of the museum or gallery has the effect of "exhibiting"[14] the cinematographic medium, as Erika Balsom writes. More obviously than it would in the space of the cinema, this pushes us (to paraphrase Marcel Duchamp), to *look* at the medium *seeing*. The artist seems to suggests here that to perceive the world, one must let oneself be affected by it, embrace its form and rhythms. The device becomes enchanted—in the sense given by Bennett of being transported—by what it films, in turn transporting the spectators.

There is something "magical" about this enchanting experience, a magic that Kim Knowles has already described as being central to expanded photochemical cinema. She locates the originality of the latter in its invitation to audiences "to revel in the magical possibilities of the medium",[15] especially in highlighting forgotten (or still to be discovered) visual pleasures that this medium can provide us. In this sense, according to Knowles, what enchants us when seeing these works is the fact that they showcase the precise physicality of the apparatus in an "effective and *affective*"[16] way.

Nominoë Collective: Simulating the Song of the World

This invitation to revel in the magical possibilities of the medium is evident in *Parallax* (2017) by the Nominoë collective in collaboration with Greek musician Michalis Moschoutis. Nominoë is less linked to the history of the artist-run film labs than Silvi Simon. While Emmanuel Lefrant worked with and in l'Abominable for several years, producing his first films and learning to manipulate the processes of development and processing there, this has not been the case since the lab moved to La Courneuve in 2010; the other members have never been associated with any lab. Nevertheless, their way of working collectively is akin to the convivial spirit that Noélie Martin discerns in these places. Each member brings their own knowledge and skills to the group, and it is by combining these contributions in a non-hierarchical way that each of their works is developed. Lefrant explains that:

> Nicolas, a professional editor who made many films around the 2000s, contributes a lot for example in terms of images. Many of our visual "motifs" come from his inventions. ... Stéphane, on the other hand, is a brilliant handyman [*bricoleur*], very skilled with his hands, who has allowed us to set up many of our cinematic "machines". He is also a sound artist and has developed many of our sound constructions. As for me, I take care of all the "lab" parts, developing all our film elements, and I also believe I contribute a lot of expertise in projection.[17]

Parallax is their most recent film performance, designed for four projectors and three screens. The 16mm projectors are placed on trolleys, which allows them to be moved on rails throughout the show, producing movements in the projected image that are directly linked to the visible movements of the machinery and of the crouching bodies of the artists who manipulate them with their hands. These movements of the projectors and of the bodies of the artists are all the more visible as they are located in front of the audience: they are part of the *dispositif*. Various multiplied, superimposed and mobile geometric shapes are projected onto the three screens simultaneously, creating a ballet of circles, rectangles, squares or flickering light beams that intersect, move away from each other or coalesce. The particular layout of the screens—one transparent, placed in front of the other two mounted perpendicularly to form a corner—creates a multitude of simultaneous viewing axes, and sometimes an impression of three-dimensional depth, producing a disorienting effect for the viewers (Fig. 23.2).

It is the soundtrack created by Michalis Moschoutis that gives the key to understanding the specific experience of enchantment in *Parallax*. Jane Bennett reminds us of the root of the word: to enchant is "to surround with song or incantation ...; to make fall under the sway of a magical refrain, to carry away on a sonorous stream".[18] Throughout the performance, Moschoutis creates a live electronic sound environment, taking as base material the sound captured by contact microphones placed on the rails and projectors, which he amplifies and digitally reworks in real time, then simultaneously broadcasts into the space via several speakers. The materiality of the *dispositif* is thus transformed into a *sound space*, a "sonorous stream" in which we are immersed and by which we are

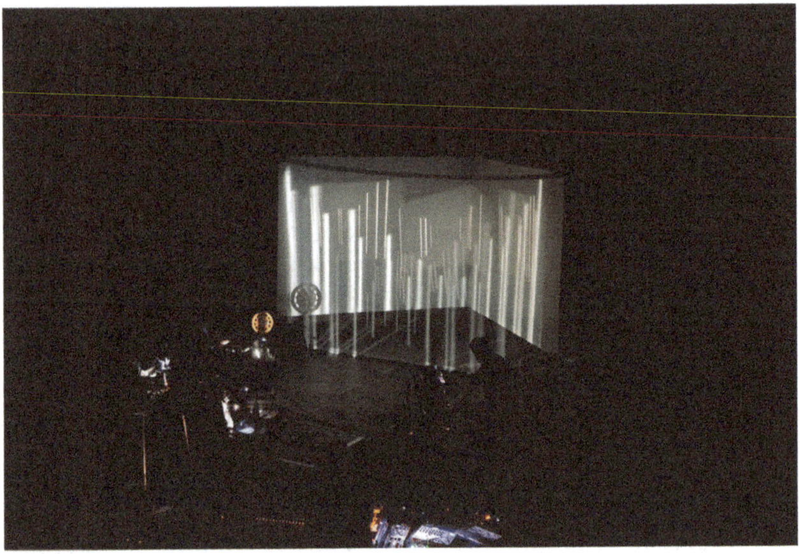

Fig. 23.2 Nominoë, *Parallaxe* (2017) © All rights reserved by the artist/ Courtesy of Light Cone

carried away. This world—kinetic and immersive, luminous and mobile—projected on the screen is physically created before our eyes by the artists who manipulate and move the projectors.

The materiality of the *dispositif* in this work imposes its own rhythm on us; we are integrated into the *dispositif* in such a way that we must adjust our perception to the parameters it sets up. This is the enchantment offered by *Parallax*: a staging of the "magical" capacities of the photochemical apparatus to make us *attune to its particular song*. Furthermore, we attune ourselves to the song of the work that is forming itself before our eyes. In his analysis of the work, Vincent Deville insists on the fact that "it is indeed Nature and the Living that we are faced with. A forest of vertical bars. A river of cubes. A primordial fold-breach [*pliure-fente originelle*]."[19] It is a "Nature" that is not only created from scratch by the *dispositif*, but which forms itself before our eyes: *natura naturans*. The Living and Nature that we see cannot be perceived as fixed, eternal entities, simply existing in a stable ontology. On the contrary, everything in the performance presents itself in the very movement of its formation, and it is to the rhythm of this self-forming movement that we attune ourselves.

Parallax invites us to attune ourselves to the song of a world in formation, but, unlike the digital works of Silvi Simon, this world is simulated, created from scratch by the cinematographic *dispositif*. Like any simulation, this one allows us to learn how to act in the outside world. Here, we learn to attune ourselves to the "song of the world".[20] Upon leaving these works, perhaps could we feel able to see the world differently, to let ourselves be re-enchanted by its song that has become inaudible to us?

Gaëlle Rouard: Entropic Rhythms

Gaëlle Rouard, co-founder of the Atelier MTK, discovered the world of experimental cinema by attending a screening organized by members of the Cellule d'Intervention Métamkine, and she said she "immediately felt the call of a vocation".[21] What attracted her most was the work in the laboratory—a form of "asceticism" that offered "almost infinite discoveries in formal research"—and the fact that she could carry out all the stages of film production herself. Rouard therefore participated in the collective management of the Atelier MTK and produced her films there for over a decade before leaving Grenoble in 2007. Her participation in various performance groups during this time—Projos Quartet with Christophe Auger, Xavier Quérel and Étienne Caire, and the Lafoxe duo with Étienne Caire—has, she said, "significantly shaped [her] gestures", teaching her to operate and consider the projector as a type of instrument. Her place of residence being far from the city these days, her environment has become her main shooting location. Living more isolated allows her a "deeper concentration", to work with less urgency and to acquire the "necessary and important distance for the appreciation of [her] films".[22]

Rouard seems to have retained from the MTK era an enchanted, almost spiritual, relationship with the filmic material and the cinematographic apparatus: she has developed an alchemical mastery of it. Her films, which she always performs live, reflect this mastery, always aiming for a "transmutation" of the material "into something vaster".[23] *M...H* (2018), one of the filmmaker's more recent works, is an enigmatic and fragmentary interpretation of Shakespeare's *Macbeth*. Perhaps more precisely, it could be thought of as the vision of *Macbeth* from the perspective of the three witches' magical cauldron that we encounter at the beginning of the play.[24] Mysterious, seemingly organic, entities, are formed and deformed on the screen: these are objects filmed by Rouard, transformed by the artist's chemical interventions on the emulsion. Other more recognizable images—mountainous landscapes, burning materials or spectral silhouettes—appear here and there, only to freeze or be distorted by the actions of the filmmaker, who manipulates the film reel during the projection or places prisms in front of the projector, diffracting the light. The entire experience is suffused with a misty, murky atmosphere, accentuated by the soundtrack composed of threatening sound effects, voice excerpts and music mostly taken from various film versions of *Macbeth*. Notably, we hear some voices from Orson Welles' version (1948), musical excerpts from Roman Polanski's film adaptation (1971), excerpts from Akira Kurosawa's *Throne of Blood* (a *chanbara* adaptation[25] from 1957) and voices from other French, Italian and German adaptations of *Macbeth*.

M...H opens with Orson Welles' voice, taken not, however, from his *Macbeth*, but from his monologue in John Huston's *Moby Dick* (1956). This suggests that the central theme of *Macbeth*—that which is created must inevitably decompose—is in fact universally applicable: the drama of *Macbeth* is also the drama of the material universe. *M...H* is therefore not so much an adaptation of the story but of the entropy personified by Macbeth. If we must learn to better attune ourselves to the "song of the world", *M...H* seems to propose that the "basso continuo"[26] of such a song is entropy, the inexorable disintegration of everything, which forces us to perceive the essential instability of forms. This is the ***transmutation*** inherent in Gaëlle Rouard's work: the story of *Macbeth* is distilled, then re-presented as an experience of the essential instability of matter and, more broadly, of the world. The work thus synthesizes the two approaches to enchantment that we have previously discussed, as the film medium is made to perform the entropic rhythm of the world, transforming the simulated world of *Macbeth* into a sensory experience for the viewers.

In 1970, Gene Youngblood already saw in expanded cinema a means of extending consciousness, allowing us to better fight against our own "disenchantment and alienation".[27] The works cited here—all emblematic of contemporary expanded cinema in France—bring this expansion of consciousness back to non-alienated, artisanal and convivial modes of creation, of which artist-run film labs are a symbol. Working in a convivial fashion produces new, magical relationships with the materiality of the apparatus. In these films, these relationships are used in different ways to try to attune us with the song of the world, channelling the expansion of perception spoken of by Youngblood towards a properly ecological experience. Thus, an *ecopoetics of re-enchantment* is understood here in the precise sense of an articulation between convivial *making* and a *perceiving* that attunes us to the song of the world: *non-alienated poiesis, enchanted aesthesis*.

Notes

1. Pip Chodorov, "The Birth of a Labo: L'Abominable, MTK et l'histoire du réseau des laboratoires indépendants en France", in Nicole Brenez and Christian Lebrat (eds), *Jeune, dure et pure ! Une histoire du cinema d'avant-garde et expérimental en France*, Paris, Cinémathèque Française / Mazotta, 2001, p. 519. Unless otherwise noted, all translations from original French are by the author.
2. Blick, "Attention la littérature: Pour une nouvelle genèse du cinéma", *L'Ébouillanté*, no. 15, June 1999, p. 8.
3. *Ibid.*, p. 18.
4. Dominique Noguez, *Éloge du cinéma expérimental* [3rd ed.], Paris, Paris Expérimental, 2010, p. 299.
5. Akeel Bilgrami, *Secularism, Identity and Enchantment*, Cambridge, Mass., Harvard University Press, 2014.
6. Silvia Federici, *Re-Enchanting The World: Feminism and the Politics of the Common*, Oakland, PM Press, 2019.
7. Nicolas Rey, "Enjeux contemporains des laboratoires cinématographiques de cinéastes", *Trafic*, no. 108, winter 2018, p. 176.
8. Bernard Stiegler, *The Re-Enchantment of the World. The Value of Spirit Against Industrial Populism*, London, Bloomsbury, 2014 [2006].
9. Noélie Martin, *Ethnographie d'une pratique filmique actuelle: la fabrication des émulsions artisanales* [Master's thesis], Paris, École des Hautes Études en Sciences Sociales, 2016, p. 103.
10. See Bénédicte Meillon, "What Matters Sings: Ecopoetics of Reenchantment", in Meillon, B. (ed.), *Dwellings of Enchantment, Writing and Reenchanting the Earth*, London, Lexington Books, 2021, pp. 1–20.
11. Interview with Silvi Simon, 26 October 2021.
12. Ivan Illich, *Tools for Conviviality*, New York, Harper & Row, 1973, p. xxiv.

13. Jane Bennett, *The Enchantment of Modern Life: Attachments, Crossings and Ethics*, Princeton, Princeton University Press, 2001, p. 5.
14. Erika Balsom, *Exhibiting Cinema in Contemporary Art*, Amsterdam, Amsterdam University Press, 2013, pp. 13–15.
15. Kim Knowles, *Experimental Film and Photochemical Practices*, Cham, Palgrave Macmillan, 2020, p. 208.
16. *Ibid.*, p. 212.
17. Interview with Emmanuel Lefrant, 2 November 2021.
18. Jane Bennett, *The Enchantment of Modern Life*, *op. cit.*, p. 6.
19. Vincent Deville, "Incidence du changement de position de l'observateur sur l'observation d'un objet: *Parallaxe* de Nominoë", *Cahier Louis-Lumière*, no. 12, "Arts filmiques et expérimentations optiques contemporaines", 2018, p. 163.
20. Jean Giono, *The Song of the World* [1934], trans. Henri Fluchère and Geoffrey Myers, New York, The Viking Press, 1937.
21. Interview with Gaëlle Rouard, 3 November 2021.
22. *Ibid.*
23. *Ibid.*
24. Agnès Perrais and Sébastien Ronceray interpret *M...H* in a similar way: Agnès Perrais and Sébastien Ronceray, "*M...H*, Le chaudron des sorcières, Un film de Gaëlle Rouard", *La Revue documentaire*, no. 31, "Films, textes, textures", July 2017, pp. 173–186.
25. *Chanbara*: a genre of Japanese samurai film.
26. A musical term, characteristic of baroque music, which denotes the part played by a low-pitched instrument and which provides the harmonic structure of the piece. In the 1920s, filmmakers Viking Eggeling and Hans Richter took from J.W. Goethe the idea that all art should find its "basso continuo", and attempted to find that of cinema. Gilles Deleuze proposed in the 1980s that Leibniz's writings on the harmony of the universe integrate baroque notions such as the basso continuo, among others.
27. Gene Youngblood, *Expanded Cinema*, New York, P. Dutton & Co., 1970, p. 47.

The manufacturer's authorised representative in the EU is Springer Nature Customer Service Centre GmbH, Europaplatz 3, 69115 Heidelberg, Germany. If you have any concerns regarding our products, please contact ProductSafety@springernature.com

Printed and bound by CPI Group (UK) Ltd, Croydon, CR0 4YY

16/09/2026

02217783-0004